MW01195492

THE FAIRNESS INSTINCT

THE FAIRNESS INSTINCT

THE ROBIN HOOD MENTALITY AND OUR BIOLOGICAL NATURE

L. SUN

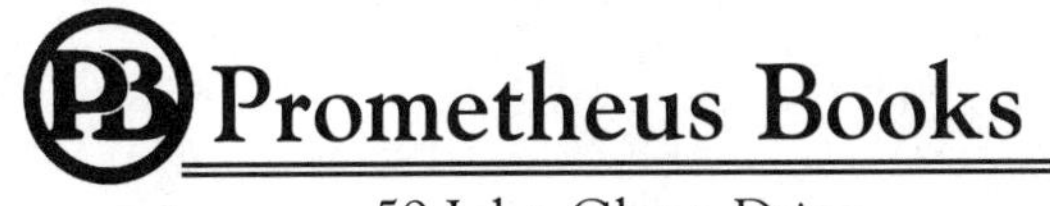

59 John Glenn Drive
Amherst, New York 14228–2119

Published 2013 by Prometheus Books

Cover design by Jacqueline Nasso Cooke
Cover image © gasa/Shutterstock.com

Inquiries should be addressed to
Prometheus Books
59 John Glenn Drive
Amherst, New York 14228–2119
VOICE: 716–691–0133
FAX: 716–691–0137
WWW.PROMETHEUSBOOKS.COM

17 16 15 14 13 5 4 3 2 1

Library of Congress Cataloging-in-Publication Data

Sun, Lixing.
The fairness instinct : the Robin Hood mentality and our biological nature / by Lixing Sun.
pages cm
Includes bibliographical references and index.
ISBN 978-1-61614-847-8 (hardcover)
ISBN 978-1-61614-848-5 (ebook)
1. Fairness. 2. Moral development. 3. Psychobiology. 4. Evolutionary psychology. I. Title.

BJ1533.F2S86 2013
179--dc23

2013022362

Printed in the United States of America

For Crystal, Shine, and Orien,

from whom I have learned the best practical knowledge about fairness.

CONTENTS

Part 1

THE ROOTS OF THE ROBIN HOOD MENTALITY

We love Cinderella stories, cheer for underdogs, hate cheaters and traitors, punish bullies and evildoers, cherish rags-to-riches yarns, wish heroes get the girls and live "happily ever after," and pray for divine intervention after massive disasters—earthquakes, hurricanes, tsunamis, terrorist attacks. These are all shoots—expressions even—of our fairness instinct. Fairness is an overarching topic that has intrigued philosophers and social thinkers in both Eastern and Western societies for millennia. Yet only in recent decades have its biological roots, as a part of our DNA, come to light. Never before have we had such a clear understanding about how fairness works and why it is so intimately involved in our lives and society.

Chapter 1

INTRODUCTION, OR, WHO IS ROBIN HOOD?

The "person" in *Time* magazine's 2011 "Person of the Year" is "the protester." It is the tenth time since its founding in 1927 that *Time* has recognized a class of people instead of an individual. From the Arab Spring to the Occupy Wall Street (OWS) movements, from Russia to India, the protesters in 2011 around the globe shared two commonalties: Facebook and a desire for fairness—the new and the old. The new social-networking tool made manifesting the old fairness sentiment all the more rapid, all the more contagious, and all the more potent.

Whether at home in the United States or abroad, these protest movements largely hinge on a widespread disaffection that stems from what many perceive as growing political, social, and economic inequalities. In many ways, the OWS movements epitomized these global convulsions provoked by the public ire over the corporate bailouts, the widening gap between rich and poor, and the basic unfairness in tax policies, the last of which continues to be a divisive issue in America. While the "Occupy Wall Street" slogan seems to have slipped from parlance after the 2012 election season, the "99%" slogan continues to gain in potency.

Contrast this to just four years earlier, how mute such rhetoric was in the election of 2008. Even after the massive Wall Street sell-off in September 2008, the likes of which hadn't been seen since the Great Depression, Republican John McCain was still vilifying Democrat Barack Obama as a modern-day Robin Hood for Obama's proposal to raise the income tax on the rich. Obama, McCain warned, wanted "socialism" and would go so far as to wage "class warfare" to "spread the wealth." McCain's pitch was apparently based on his take that Americans disliked playing Robin Hood. Such a public mentality—even if true at the time—would soon retreat.

In March 2009, the stock market continued its downward spiral, and the resulting financial tsunami wiped out American jobs en masse. In the meantime, executives at AIG were pocketing large annual bonuses—some $165 million in total. Americans—liberals and conservatives, Democrats and Republicans—were furious. Only a few months earlier, the insurance behemoth was on the verge of bankruptcy. AIG would have gone belly-up if not for the timely infusion of $182 billion in bailout money, courtesy of American taxpayers. For the American public, rewarding fat bonuses to rich executives at a difficult time seemed equivalent to robbing people of their possessions. Amid a national outcry, the US Congress on March 19 passed an urgent bill to levy a staggering 90 percent tax on all similar payouts, only days after the news about the bonuses broke. Considering that Congress had been partially paralyzed by partisanship for years, it was surprisingly efficient in passing this special bill, intended to seize the bulk of the money from the bonus recipients. Among the supporters of the bill were eighty-five Republicans, who would otherwise hold lower taxes among their core conservative principles. Due to mounting public pressure, their decision was more about political survival than political principle, for the time being.

Company executives at AIG felt victimized. Despite the fact that 80 percent of AIG was owned by the government, the bonuses at the time accounted for only 0.076 percent of the company's value, a minute proportion, making it hard to claim they were entirely from taxpayers. Also, the bonus agreement had been set up well before the bailout. It was a contract that should, in principle, be honored. Furthermore, AIG was not the only company handing out extravagant bonuses, compensations, and golden parachutes to executives. All major financial firms either taken over by the government or resuscitated by TARP moneys—Fannie Mae, Freddie Mac, Goldman Sachs, Citibank, Bank of America, Wells Fargo—had done the same thing.

For AIG, it was unfortunate to be situated in the bull's-eye of public fury, but the political storm left a trail of questions in its wake. Why did bonuses and executive compensations suddenly matter so much in the psyche of the

American public? Why was the voice against lavish pay packages for company executives so feeble before the election? Why, in a short four-month period, was there a sea change in public opinion? What was the powerful force behind this dramatic turn of events? The key to these mind-boggling questions lies in our Robin Hood mentality—a metaphor for our sense of fairness.

Who, then, is Robin Hood? The answer seems obvious for most of us. Fleshed out in ballads and folklores, Robin Hood is a legendary hero who robs the rich and gives to the poor. He is a savior for the poor and powerless. He opposes tyranny and fights against rich clerics and officials who are hated for their abysmal greed and wicked corruption. The legendary Robin Hood, agrees historian Maurice Keen, "displays many of the characteristics traditionally associated with fictional knightly heroes—courage, courtesy, loyalty, generosity, a free and open bearing."[1]

Robin Hood is the hero of countless poems, songs, plays, novels, and comic books. Movies made Robin Hood all the more vivid: forty-nine Robin Hood films and TV series—nearly one in every two years, on average—have been produced since 1908.[2] In the age of information, Robin Hood has gained a new life on the Internet. The University of Rochester, for instance, maintains a special website dedicated to scholarly studies of literature, history, and folklore of Robin Hood.[3]

But there is a dilemma. A popular hero is usually well known for his name, birthright, and social stature. Yet exactly the opposite is true for Robin Hood. In fact, little is known as to who he really was beyond the stories in pop-cultural products. Even the "basic" facts about him are controversial at best. Is Robin Hood even the legendary hero's true name? Some believe the name Robin was derived from, among others, Robyn or Robert, both common in medieval England. The last name might be Hode or Hude, in addition to Hood. Robin Hood might simply be a generic name for fugitives and outlaws, such as Hereward the Wake, Eustace the Monk, Fulk Fitzwarin, and William Wallace, who all gained some fame among English folks at the time.[4]

Another controversy concerns the time Robin Hood lived. John Major, a sixteenth-century Scottish historian, believes that Robin Hood was active in 1193 or 1194, when King Richard's brother, John, attempted a coup against Richard. *Wikipedia*, the open-source encyclopedic website, refers to the earliest mentioning of Robin Hood in 1228.[5] In the absence of solid evidence, the dates of 1193 and 1228 could each be correct, depending on Robin Hood's age. Thomas Gale, dean of York between 1697 and 1702, thought Robin Hood died "on the 24 Kalends of December 1247."[6] Keen believes "it is a very reasonable conjecture" that Robin Hood was still alive in 1266.[7] "Considering the silence of chroniclers and poets alike before that time, it would seem very unlikely that he could have lived earlier than, say, the first quarter of the thirteenth century."[8]

The third black hole surrounding Robin Hood's dubious identity is his social status. Earlier ballads put him as a yeoman. In preindustrial England, yeomen were commoners who owned small parcels of land.[9] They sat above peasants in the English social hierarchy. Even so, there is some ambiguity. The word *yeoman*, according to historian John Bellamy, can refer to "both household servant and freeman."[10] In the fourteenth century, it referred to artisans as well. An artisan, by definition, did not belong to the landowning class. So Robin Hood's exact social role isn't crystal clear, and folklores, in historian J. C. Holt's assessment, have done little but muddy the water. "At his first appearance Robin was a yeoman. He was then turned into a nobleman unjustly deprived of his inheritance, later into an Englishman protecting his native countrymen from the domination of the Normans, and finally into a social rebel who, in the peasant's struggle against the grasping landlord, retaliates against the person and property of the oppressor."[11]

Even if Robin Hood's social status were beyond doubt, a yeoman, notices Keen, would be the least exciting person to ascend to the status of legendary hero. "Robin Hood the yeoman is not the kind of figure to catch the limelight of medieval historical writing. He has no part to play on the grand political stage; he is a forest robber of humble origins and his cause has to do with the conditions of the everyday social world, not with the melodramatic conspiracies which troubled the sleep of kings."[12] Quite to the contrary, heroes typically emerge from the lower or the upper class of society, but rarely from the middle. Indeed, gallant aristocrats aside, many historic heroes—such as

Spartacus, the Roman slave, and William Tell, the Swiss peasant rebel—rose from the downtrodden or the oppressed in Western societies. Sitting snugly on the middle rung of a social hierarchy, yeomen were probably the least inspiring and colorful, simply because they were too common, too plain, or too little motivated to do something dramatic. They lacked the theatrical aroma necessary for delicious yarns.

Unsurprisingly, Robin Hood was elevated to a noble in later versions of his legends. Playwright Anthony Munday named Robin Hood Earl of Huntington in 1598.[13] And, in 1746, William Stukeley, an officious Royal Society fellow, cooked up a spurious aristocrat pedigree for Robin Hood.[14] Such fudging paved the way for legitimizing the romance between Robin Hood and Maid Marian, the sister of King Richard, in a time when status mismatch was a stiff taboo for marriage in English society. With love and hatred peppering up an otherwise dull yeoman's life, we now have a chivalric hero.

Last but not least, how did Robin Hood become a lawless fugitive? One theory suggests that he was forced to become an outlaw by King Richard's evil brother John while Richard was away, leading the eastward march during the Third Crusade. Alternatively, Robin Hood was "indicted out of malice" by the sheriff of Nottinghamshire and forced to live in the woods. The sheriff's malevolence goaded Robin Hood to serve justice with his own hands. Though a popular and seemingly reasonable account, little historical evidence gives it any support. Even the bedeviled sheriff had no real match among those who ruled the area at the time. "Robin's tale . . . is imprecise," Holt concedes. "He is an outlaw, but no one explains why. He is in conflict with the sheriff but no reason is ever given; it is simply that the sheriff represents the law and Robin stands outside it. His story is less committed to immediate circumstances."[15]

But regardless of the dearth of evidence regarding whether he is a real person,[16] Robin Hood continues to survive as a shining hero in traditional and modern media in those societies thick with English culture. Apart from Jesus Christ, few other fabled figures have enjoyed such broad, timeless appeal and influence. Why has Robin Hood been so popular for so long? Paradoxically, part of the answer lies beyond medieval England.

Hardly confined to English folklore, incarnations of Robin Hood can be found in many other Western nations and societies, including "Rob Roy" MacGregor in Scotland, Louis-Dominique Bourguignon and Robert Mandrin in France, Johannes Bückler and Matthias Weber in Germany, Diego Corrientes in Spain, Angelo Duca in Italy, Stenka Razin and Emel'yan Pugachev in Russia, Juro Janošik in Hungary, Chucho El Roto and Jesús Malverde in Mexico, to name just a few.[17] All of them were credible historical figures. Like Robin Hood, they were outlaws and rebels who robbed the rich and gave to the poor. Like their English parallel, they have been hailed as folk heroes of some sort. Were they copycats of the English Robin Hood?

Eastern folklore is not short of versions of Robin Hood. In China, Korea, Japan, and other East and Southeast Asian countries, Robin Hood–like characters masquerade as martial artists or samurais with insuperable fighting skills. They, too, rob the rich and help the poor as their way of serving justice. In countless stories, ballads, legends, and fairy tales, these martial artists come to rescue people from local toughs, bullies, and villains, who are stereotypically the rich and corrupt. Such chivalric stories are so popular in China today that they represent a unique literary genre of their own in books, movies, and TV series, usually with predictable plots and endings. Apparently banal to the creative mind, they nonetheless claim large numbers of readers and viewers, enjoying a substantial share in the media and entertainment market. Westerners can get a glimpse of this genre in movies ranging from those starring Bruce Lee to more recent Chinese ones such as *Crouching Tiger, Hidden Dragon* and *Hero*, which all reflect a general theme otherwise completely missing in the West—if not for the sake of Robin Hood. How can we resolve the dilemma of why there have been so many different versions of Robin Hood in so many cultures for so long, if Robin Hood is only a fictional figure? Is this a mere coincidence between Western and Eastern cultures?

Today, while throngs of Chinese are fixated upon their kung-fu Robin Hoods, Americans are engrossed in Robin Hoods of their own creation, regardless of vast differences in history and culture. Revenge stories have dominated the silver screen with such classic films as *Ben-Hur* (1959), the

Godfather trilogy, the *Death Wish* series, *Carrie* (1976), *Unforgiven* (1992), *Braveheart* (1995), and *Gladiator* (2000), to name only a few. The stereotypical heroes in American Western movies epitomized by Gary Cooper, John Wayne, and Clint Eastwood are largely variants of Robin Hood. And, in their upgraded versions, cowboys are replaced by testosterone-charged idols in the bodies of Sylvester Stallone, Arnold Schwarzenegger, Chuck Norris, Steven Segal, and other macho actors. As though the musculature of these invincible men is not impressive enough, Hollywood has resorted to supernatural heroes—Superman, Spider-Man, Batman—who possess unearthly prowess to resolve problems insurmountable by our own species. What can we make of the popularity of these movies and their protagonists?

"[The Robin Hood] legend is about justice," observed Holt decades ago. "Robin is . . . an embodiment of honour and an agent of retribution. . . . Robin also foreshadows the world of superman and the comic strip."[18] The broad appeal and off-the-chart box-office success of many of these surreal movies can hardly be entirely explained by their artistic creativity and dazzling special effect; it also lies in the themes of the yarns that resonate in the viewers: a desire for fairness and justice.

A secular value of universal appeal, fairness is omnipresent and often sacred, to a degree that major religions—Buddhism, Judaism, Christianity, Islam—claim it as a religious virtue. Most, if not all, societies—ancient or modern, tribal or industrial—have some notion concerning equality and fairness in economic, social, and political status.

Probably *the* most essential rule in social engagement, fairness has shaped human relationships, molded human societies, and directed the course of civilization. It governs virtually all aspects of our society, from economics, politics, education, and military organization to sports and entertainment. Its long arms grip issues as small as division of family chores between a couple, gift exchanges between friends, and daily interactions with coworkers. It also underlies major social, economic, and political issues such as taxation, gender equality, racial relationships, and international affairs.

Furthermore, fairness is the foundation for justice—the most important moral principle in human societies. For French anarchist Pierre-Joseph Proudhon, government was disposable; justice was not:

> Justice, under various names, governs the world—nature and humanity, science and conscience, logic and morals, political economy, politics, history, literature and art. Justice is that which is most primitive in the human soul, most fundamental in society, most sacred among ideas, and what the masses demand today with most ardour. It is the essence of religions and at the same time the form of reason, the secret object of faith, and the beginning, middle and end of knowledge. What can be imagined more universal, more strong, more complete than justice?[19]

Indeed, fairness underpins many landmark events in both Eastern and Western societies. In the West, we have adopted the blindfolded Roman goddess Justitia—Lady Justice—as the symbol of fairness and justice. The American Revolution was an uprising against the unfair taxes imposed by the British. The French Revolution of 1789 was rooted in the widening economic gap between rich and poor, as were Russian revolutions in the beginning of the twentieth century. So, too, were the sweeping Communist movements in China, Southeast Asia, Latin America, and Africa—despite their disastrous consequences. These latter examples illustrate that our relentless quest for equality, fairness, and justice may lead to outcomes that are not always desirable. In fact, the sense of fairness can lead to great friendship, partnership, long-term collaboration, team spirit, and global peace, on one hand, and suspicion, ill will, anger, retaliation, feuds, and mass violence, on the other.

What are fairness and justice? The answer varies. Some use the two words interchangeably, whereas others make clear distinctions. Justice, in the most generic term, "concerns what people are due," according to a recent textbook.[20] A more specific definition is given by political philosopher Michael Sandel, who simply stamps it in the title of his book: *Justice: What's the Right Thing to Do?* Justice guides our moral judgments, decisions, and actions. As a main concern in our society, the issue of justice seeps into all spheres of our social lives: political, economic, social, legal. Distributive justice, for instance, concerns how key resources and opportunities are allocated in society. It has played a major role in all human societies and will thus be the focus of this book.

For time immemorial people have been seeking to define justice and the ways to implement it to create a better society. Such intellectual pursuits are evident in the texts of Lao Zi and Kong Zi (Confucius) in the East, and the Talmud, as well as the writings of Aristotle and Plato, in the West. In fact, all of Confucianism is hinged on two core values: beneficence (*ren*) and justice (*yi*), which are the backbone of the government and moral system of a society. Confucius cautions that when a profit is up for grabs one must consider whether grabbing it is the just thing to do. Early Indians, likewise, saw justice at two levels, *niti* and *nyaya*. The former "relates to organizational propriety as well as behavioural correctness," explains Nobel laureate economist Amartya Sen, whereas the latter "is concerned with what emerges and how, and in particular the lives that people are actually able to lead."[21]

In the West, meanwhile, ancient Greeks such as Theognis and Plato consider justice the sum of all virtues. It makes people act rationally and dutifully in such everyday affairs as keeping promises and paying debts. Aristotle holds that reward should be based on merit, in terms of political status.[22] In his words, "all men agree that what is just in distribution must be according to merit."[23] In a similar vein, justice, for Byzantine Emperor Justinian, "is giving to each person what is due to each."[24]

Where does justice come from? As early as the first century BCE, the Roman orator Cicero considered justice to be natural and universal for humans.[25] But almost four centuries later, Augustine believed that justice is based on values specific to Christians; that is, no justice will prevail in people who fail to put their faith in the Christian God. It is surprising that, even within the vast territory of the Roman Empire, which encompassed so many cultures, he missed the obvious fact that non-Christian societies dealt with daily issues fairly and justly. Still, Augustine's religious view was so influential that his conception of justice prevailed in the West until the thirteenth century, when Thomas Aquinas began to see justice as a more or less universal virtue that occurs in different cultures, religions, and political systems. Yet, until the seventeenth century, Aquinas's thought could do little to shake the concepts established by Augustine.[26] Immanuel Kant was a key figure in completing the paradigm shift. He affirmed that the virtue of justice has little to do with religion or culture. His take on the universal nature of justice was that each human being "exists as an end in himself and not merely as a means" and that each "has an equal right to the good things which nature has provided."[27]

Aquinas's idea about distributive justice is primarily Aristotelian, centered on merit—that is, political status—with little concern for the poor and needy. The Enlightenment movement turned this ancient view on its head; justice became an entitled right belonging to all in a society. John Locke, for instance, believes "justice gives every man a title to the product of his honest industry and the fair acquisitions of his ancestors."[28] For David Hume, "a rich man lies under a moral obligation to communicate to those in necessity a share of his superfluities."[29] In his view of distributive justice, Adam Smith also addresses the needs of others, in addition to a person's merit. Although he is not overly concerned about the misery of the poor, he does put justice within the radius of the Christian social virtues. Both Hume and Smith advocate for property rights, which, they believe, would eventually be beneficial to the poor.[30]

In stark contrast to property rights for individuals, the concept of communal ownership of resources and wealth also enjoys a large following. Although today we tend to credit the idea to Karl Marx, its origin can be traced much further back in history. Plato, for instance, advocates it in his major work, *The Republic*. In the New Testament, Jesus lives among the poor and the apostles, knit into a cohesive community, within which "distribution was made unto every man according as he had need." This is an ideal that has been attempted by many Christian sectors, such as the Franciscan order, the Anabaptist rule (1534–1535), the communities of the Diggers (1649), the Shakers, the Oneida community, and many others, including even the notorious Temple of God founded by the cult leader James Jones.[31] Thomas More's epic *Utopia* introduces an enticing world without private property, where members share everything. Rousseau, though a contemporary of Smith and eighty years the junior of Locke, condemns the existence of private property as the fountain of inequality, violence, and social discord. When Karl Marx penned the famous words, "from each according to his ability, to each according to his need" in *Critique of the Gotha Program* in 1875, the Communist idea of distributive justice was anything but original.

Before the Enlightenment, alleviating the misery of the poor and needy was viewed as an act of benevolence or merciful charity—a moral option rather than an obligation; it is viewed as an act of justice in the modern sense of distributive justice. Today's distributive justice hinges on meeting the welfare of people in regard to food, housing, education, healthcare, and other basic needs. All people, as human beings, deserve and are entitled to some essential goods, regardless of who they are. Hence, distributive justice, in today's sense, has little to do with merit. Only after the basic needs of the people are met does the issue of merit come up.[32]

This modern conception of justice owes a great deal to John Rawls, who advocates two principles: the equal liberty principle and the difference principle. The former mandates that all people have basic equal rights, including rights to free expression and association, and to holding and using personal property. The latter demands that, equal opportunity aside, social and economic inequalities be arranged to the greatest benefit of the least-advantaged of society.[33]

Although a persistent quest for equality and justice has placed fairness at the center of social and political movements that have profoundly transformed Western societies since the Enlightenment, an intriguing dilemma remains. On one hand, nearly all leading philosophers, including Rousseau, Hume, Kant, and Rawls, have laboriously tackled the issue. On the other, the lack of consensus on the meaning of justice has left us with heated debate on virtually every major social issue: taxes, affirmative action, government spending, gay marriage, universal healthcare, environmental protection, and immigration, to name just a few. Despite Rawls's influential ideas about what ideal justice should be, we are still far from reaching consensus on a vast array of practical issues. The problem lies in a seemingly simple fact: while a vast amount of effort has been poured into hunting for ideal principles of justice, fairness—the evolutionary underpinning for justice—has been overlooked. Compared with her more glitzy twin sister justice, the reticent fairness has gotten little attention. Even Rawls, whose name is forever tied to his advocacy for justice as fairness, never provided an explicit definition for fairness in his writing. The following paragraph appears his closest miss:

> Immediately the question arises as to how the fair terms of cooperation are specified. For example: Are they specified by an authority distinct from the persons cooperating, say, by God's law? Or are these terms recognized by everyone as fair by reference to a moral order of values, say, by rational intuition, or by reference to what some have viewed as "natural law"? Or are they settled by an agreement reached by free and equal citizens engaged in cooperation, and made in view of what they regard as their reciprocal advantage, or good? Justice as fairness adopts a form of the last answer: the fair terms of social cooperation are to be given by an agreement entered into by those engaged in it.[34]

Clearly, for Rawls, fairness is a social contract among people under fair conditions.[35] Between the lines, Rawls was conscious of differences in people's perception of fairness. This becomes more explicit for Sen, who is among the very few thinkers who actually provide a definition of fairness. Fairness, according to Sen, "can be broadly seen as a demand for impartiality" uninfluenced by "our respective vested interests, or by our personal priorities or eccentricities or prejudices."[36]

Fairness is probably the most important rule that governs our daily interactions with spouses, lovers, friends, coworkers, and neighbors. Yet conflicts may still arise regardless of how well-meaning we are and how faithfully we stick to the principle of being fair. Sometimes trivial misunderstandings in conversations, gift exchanges, and other social events can mire us in suspicion and quarrels, or even lead to the dissolution of long-term relationships. The most common reason is that what is perceived as fair by one party may be deemed unfair by the other.

Take income taxes for example. Is it fair to put people with greater income into higher tax brackets, a system known as progressive taxation, which assumes that those who earn more can afford to part with a larger proportion of their incomes? If the answer is yes, the economic advantage of those who do well is curtailed by the system, which, as some argue, is unfair. If the answer is no, why is such a policy supported by the vast majority of people, especially among "the 99%"?

Our legal systems are rife with similar controversies. Despite Rawls's repeated emphasis on "justice as fairness," people's views can still diverge markedly as to how justice should be served. For instance, can we penalize a pickpocket by death? Most would deem it far too harsh in today's democratic, industrial nations. Yet this was the reality in Dickensian Britain less than two centuries ago. What about stoning a person to death for the crime of adultery? Westerners would clearly say no, but many in conservative Muslim nations would find this punishment fair.

People perceive fairness differently in the moral dimension as well, where skewed and double standards are all too common. This is a major reason why ruthless robbers such as the James Brothers, Billy the Kid, and Bonnie and Clyde can gain the aura of modern Robin Hoods in America. The same can be said for Fidel Castro. Despite often being portrayed as a dictator in Western media, he is revered by many in Latin America as a hero with the courage to stand against the United State, the Goliath in the world. Even more remarkably, Osama bin Laden has been hailed as a hero in some Arab communities. In the wake of the 9/11 tragedy, while most people in the world were shocked, aggrieved, and saddened by the massive loss of innocent lives, there were instances of jubilant celebration in the streets of Iran, Iraq, and Palestine, and in the chat rooms and blogosphere on the Internet. In Iran, interestingly, spontaneous street celebrations aside, there were also candlelight vigils in solidarity with the United States.

So it is a fascinating and important question: why are our views of fairness and justice so diverse? Apparently, people's views vary along a variety of axes—biological, social, economic, religious, and other cultural aspects as well. Our sense of fairness, therefore, is expected to reflect these differences. Yet, traditionally, fairness has been approached as an ideological issue. Philosophers and social thinkers, while diligent in pursuit of universal moral standards for a perfect society, tend to overlook the vast differences in our individual senses of fairness. Aware of such peculiarities, Sen, Rawls, and many others have come close to discovering the rich lode underneath the variations in fairness perception. But, somehow, they let go of the golden opportunity.

Moral philosophy and moral psychology are distinct territories in our quest for fairness and justice.[37] Over two centuries ago, Rousseau had already made a distinction between our theoretical and practical pursuits by pointing to a simple fact: "we shall not be obliged to make man a philosopher before he is a man."[38] The common confusion between justice as abstract ideology and fairness as concrete perception makes philosopher Robert Solomon lament, "We have over-intellectualized our feelings about justice, with the result that our feelings have become as confused as our theories, if indeed they have not been eclipsed by them."[39]

Indeed, subconscious intuitions, rather than conscious moral reasoning,[40] weigh heavily in judging fairness in everyday life. More often than not, it is our intuitive emotions, not reasoned philosophies, that compel us to take action. That's why political debates can get emotional and, at times, stray off the course of civility. Emotions aside, how often do we see fairness issues settled by debate? Yet how often is the diversity of our fairness opinions given serious thought? This familiar experience speaks for the importance of understanding fairness in terms of behavior and psychology—a vital aspect missing from traditional approaches to fairness and justice issues.

Fortunately, recent studies in biology, psychology, anthropology, and economics have unveiled a much clearer and richer image of the nature of fairness. In addition to undergirding our concepts of justice, fairness indeed has its profound biological origins, rooted in our genes. Under this new light, our disagreements regarding what fairness is reflect as much about our biological nature as they do about our culture. That's why our attempts to reach a consensus on fairness are so often doomed to become wild goose chases. Such vain efforts would have been saved had we seen the gaping fault line between fairness as cultural ideology and fairness as biological instinct.

After we shift our perspective on fairness from ideology to biopsychology, the Robin Hood enigma described earlier is an enigma no more. It is clearly no coincidence that Robin Hoods, albeit with different names, exist in both Eastern and Western societies. Robin Hood echoes a common theme in human history—people yearn for justice when oppressed by a system of law and unfair social institutions that are stacked against them. As long as our social system is imperfect—if there is such a thing as a perfect one—we question its fairness and legitimacy; we want to change it. It is this human desire that makes peoples across the world crave Robin Hoods, heroes who will bring fairness and justice to their lives.

This universal desire is the creative force behind Robin Hood stories. "The identity of the man matters less than the persistence of the legend," remarks Holt. As societies, values, and times change, so do the stories. "New tales were added to the story. New characters were introduced to the plot. Fresh historical contexts were invented. Minor features of the older tales were expanded into major themes; important elements in the earlier tales were later jettisoned. The legend snowballed, collecting fragments of other stories as it rolled along. The central character was repeatedly remodeled."[41]

We are now in a good position to take a shot at the question of whether Robin Hood is real. Several scholars have offered their insights. "[Robin Hood's] struggle is part of man's ageless war against abuse," writes Keen, "and his story is more vivid and human than that of any participant in the struggles of, say, the Norman conquest or the reign of King John, because it is more universal."[42] He then elaborates:

> The outlaws of legend, if they undeniably belong to the world of mortal men, have equally undeniably been enlarged and romanticized to more than life-size in the minds of the people who heard and loved the ballads which recounted their deeds. Indeed, though their activities are corporeal enough, their stories do not have quite the ring of true history. The theme of the righting of wrong done, of the lightening of the load of the peasant and the defeat of social injustice, which invests all the outlaws' acts with a king of chivalry, does not really belong to the history of highway robbery. No robber ever could have lived as the Robin Hood of legends did; he is an ideal, not an actual figure.[43]

Along the same logic, it makes sense that Robin Hood is a yeoman. "His story belongs essentially to the common people, whose hero par excellence he was."[44] Behind Robin Hood and his stories are ordinary people who shared the desire for fairness and justice. "The characteristic plot of the Robin Hood ballads, and indeed of all the later outlaw stories, is very simple. It is a tale in which wicked men meet a merited downfall, and the innocent and the unfortunate are relieved and rewarded. As the wicked are always the rich and powerful and the innocent the victims of poverty and misfortune, they may be said to be in essence stories of social justice."[45]

This also explains why the popular appeal of the hero never wanes. In fact, Robin Hood continues to capture our admiration and fascination. As a case in point, adding to the popularity of Kevin Costner's 1991 movie *Robin Hood*, Bryan Adams romanticized the hero with his record-breaking song "Everything I Do, I Do It for You." It stayed at the top of the charts in the United States, Britain, Germany, France, Australia, and many other countries in the 1990s. Just nine years later, another big Hollywood star, Russell Crowe, became the lead actor in a movie with the same title and almost exactly the same plot. The charm of Robin Hood is indeed timeless.

There is a moral equivocation, though. When we savor Robin Hood stories or read them to our children, how many of us are aware that Robin Hood is a lawless poacher, an incorrigible robber, a shameless extorter, a ruthless bandit kingpin, a heartless kidnapper, and a cold-blooded murderer? Robin Hood prefers a society without law. He treats the rich as the wicked, as if wealth necessarily corrupts and transforms a person from good to evil. Do we really welcome such a heroic savior in our society? If not, why do we still cherish the very idea of robbing the rich and giving to the poor, regardless of Robin Hood's brutality and savagery?

Perhaps something deep within our psyche shares Robin Hood's version of fairness and justice. As we watch or read his story, we become the embodiment of Robin Hood ourselves. His blood flows in us; his emotions infect us; his primeval sense of justice lives in us. Such a subconscious desire for justice is the emotion behind the appeal of many books and movies; it is a universal theme. When justice is unserved, we feel vexed and unsettled. We experience a compulsive urge to set things right.

One outlet for this urge is spectator sports. Inside a noisy stadium, we can cheer for the scrappy underdog (David with his sling) and boo the towering

Goliath. Excepting diehard fans, who are always loyal to their team, we tend to despise the powerful franchises, simply for their star-studded rosters backed by big bucks. Among those that best fit this latter role are the New York Yankees and the Los Angeles Lakers. Part of their unwitting and unwilling duty is to entertain us by satisfying our desire to see giants fall.

In a more peaceful yet no less forceful realm, Cinderella stories and rags-to-riches tales constitute a bulk of all films and TV programs. From *My Fair Lady* to *Pretty Woman*, from *American Idol* to the overnight fame of Susan Boyle, they reflect our compassion for the weak, good, kind, and hardworking people whose lives are, in our view of a just world, unnecessarily hard. We empathize with them; we love to see them succeed and thrive. We hope that true lovers get married, poor and kindhearted people win the lottery, bullies are vanquished, and villains suffer their just deserts. And, of course, the hero always gets the girl, and, from then on, they live happily ever after, a trite finale whose appeal never seems to fade. Yet underneath the worn-out veneer of these stories lies our deeply engrained desire for fairness. In this sense, we are all Robin Hoods.

While we may be disappointed by the lack of substantive evidence for the reality behind the Robin Hood legend, it appears to matter little. We have stumbled onto something much larger than the legendary folk hero—the Robin Hood who lives inside us all. Our brief venture as historical detective has paid off richly, albeit in an unexpected way.

While leaving the search for Robin Hood's true identity to professional historians, this book invites you to tour the scientific wonderland of fairness and its social ramifications. It attempts to answer such questions as where our sense of fairness comes from, which emotions and actions emerge from our fairness instinct, how powerful and prevalent fairness can be in our social interactions, and how we can harness it to improve human relationships at all levels.

Let's start our tour. The first stop is the evolutionary genesis of our fairness sense.

Chapter 2

THE MAKING OF ROBIN HOOD

Up until the 1970s, why fairness is universally valued in human societies remained an unanswered question, despite persistent efforts by generations of philosophers. Unbeknownst to most social thinkers of the time, a silent revolution was brewing in the scientific community. As biologists started examining animal social behavior through the lens of evolution, they opened an inquiry into many human behaviors we had taken for granted—the origins of which had stumped philosophers for centuries. In 1987, Richard Alexander, a University of Michigan biologist, triggered an intellectual storm with his work *The Biology of Moral Systems*, which marked the entry of science in the debate over the nature of morality. The motive for his thought-provoking book, Alexander wrote, was the impasse in the traditional pursuit of moral issues:

> I believe that something crucial has been missing from all of the great debates of history, among philosophers, politicians, theologians, and thinkers from other and diverse backgrounds, on the issues of morality, ethics, justice, right and wrong. Why have the greatest minds throughout history left such questions seemingly as unresolved as ever? . . . Part of the answer is that those who have tried to analyze morality have failed to treat the human traits that underlie moral behavior as outcomes of evolution—as outcomes of the process, dominated by natural selection, that forms the organizing principle of modern biology.[1]

He went on to lay out the argument that morality is an evolved set of behavioral rules and norms aimed at resolving a central issue in social living: conflict of interest.

Alexander's thesis, as provocative as it was, was still in need of solid evidence because, at the time, not many studies had been specifically designed and conducted to examine moral behavior in animals. Fortunately, soon after the publication of Alexander's work, the gap between his theories and the evidence began to narrow, thanks to a flurry of new findings in social animals such as dogs, wolves, and primates. Still, after well over a decade of research, decisive evidence for any prototypes of morality in animals—including fairness and justice—was still missing.

The September 2003 publication of the short paper "Monkeys Reject Unequal Pay," appearing in the science journal *Nature*, immediately caught people's attention. Two Emory University researchers, Sarah Brosnan and Frans de Waal, reported a surprising discovery in the brown capuchin monkey native to the jungles of Central and South America.

The monkeys in Brosnan and de Waal's study were trained to use stones and pebbles as tokens in exchange for food rewards from researchers. To test whether the monkeys possessed a sense of fairness, Brosnan and de Waal designed an innovative experiment. They paired up the monkeys and gave each of them a rock as token money. If a monkey returned the rock, it was rewarded with either a grape, which they considered a highly valuable food item, or a slice of cucumber, which was not so desirable. The researchers found that when two monkeys in a pair were given the same rewards for rock tokens—either grapes or slices of cucumber—both were willing to participate in the exchange game. However, when the rewards were unequal—that is, one monkey got a grape while the other received a slice of cucumber—the shortchanged monkey often refused to cooperate in the exchange. Some even threw the cucumber slices back at the researchers in protest. Robin Hood would have been amused that his vision of justice concurs with that of capuchin monkeys.

The study proved significant. It settled the long-standing question of whether animals have a sense of fairness. Additionally, it added a solid piece of scientific evidence to the age-old philosophical contention that the sense of fairness is a universal human trait. The first conclusion is evident. The second and more crucial point, however, may take a bit of explaining. After all, how can a universal moral sense in humans get support from a simplistic prototype found in one of our distant cousins? Unknown to Aquinas and many of his theoretical compatriots, the answer lies in a philosophical rule called parsimony, a logical principle that guides us to choose the best explanation when faced with several choices. The principle holds that when many alternative theories exist, the best is the one that works for the greatest number of situations yet is based on the fewest assumptions. Also known as the law of parsimony, economy, or succinctness,[2] the principle has a colorful alias, Occam's razor. The name is derived from the fourteenth-century English theologian William of Ockham. His formulation was that "entities must not be multiplied beyond necessity."[3] Plainly stated, the simplest explanation is the best. "Razor" here is likened to a tool to cut away any undue contingencies.

Though not without flaws, the parsimony principle has a solid footing in probability theory. It tells us that, everything being equal, the more specific a story is, the less likely it is true. This is precisely what we experience when using an Internet search engine. The more specifications a query has, the fewer results will pop up. In other words, a possible event, when chained to more conditions, becomes less probable. For example, it is quite easy to pick the most likely event out of the following three as the condition of time becomes increasingly specific:

1. The world will end.
2. The world will end in the 2180s.
3. The world will end at 12:32 PM, May 4, 2188.

Thus, probability theory not only illustrates the legitimacy of Occam's razor, it also bolsters William of Ockham's esthetic credo, shared with a multitude of scholars since the heyday of the Athenian civilization: simplicity is elegance.[4]

While there may be many explanations for a sense of fairness found in both humans and their primate cousins, the principle of parsimony tells us that the simplest answer is the most likely. In this case, if a trait is seen in both humans and animals that are closely related to humans, say chimps, the most likely—most parsimonious—explanation is that the trait had already existed before humans and chimps parted in their evolutionary lineages. This historical scenario, known in Darwin's words as "common descent," elucidates why we can better understand ourselves by studying a broad range of animals. Common descent, backed by the parsimony principle, allows us to look into human nature in ways we had barely thought possible until recently. We can do this by comparing aspects of behavior and psychology—such as aggression, territoriality, emotion, personality, and mate preference—between ourselves and animals. Indeed, recent studies in monkeys and apes have hugely advanced our knowledge about a wide range of behaviors: consciousness, tool use, language, and culture, all of which were widely—but wrongly—conceived of as uniquely human only a few decades ago. Brosnan and de Waal's discovery of fairness in the capuchin monkey has further blurred the line between humans and other animals in a new territory: the sphere of morality.

Based on the parsimony principle, the evolutionary arrival of fairness might have occurred before the primate lineage branched off from other mammals tens of millions of years ago.[5] The question for us now is why fairness popped up in the first place. What might have spurred the emergence and evolution of fairness in animals? And what function does fairness serve? Part of the key to answering these questions lies in monkeys like the capuchin. But monkeys are complex animals—not only are they highly intelligent; they are also highly social. Which of the two—mental capacity or close-knit social life—hosts the evolutionary wellspring of fairness?

Unfairness evokes strong and spontaneous emotional reactions—ranging from mild discontent and envy to resentment and outright hatred—in humans, and possibly in other primates as well. These emotions goad us to take action in an attempt to curb the perceived injustice. It appears fairness, in this chain of responses, involves complex cognitive processes. In the world of Brosnan and de Waal's capuchins, a monkey has to be able to compare its reward with what its peer gets for the same "service"—exchanging rocks with humans. For the monkey, the reward of its peer sets the bar for its expected gain. The gap between the actual and expected is then used in judging whether the reward is acceptable or not—that is, whether the reward is fair or unfair. All of these steps are hinged on a basic sense of quantity, by which the magnitude, if not the precise amount, of the deviation between the actual and expected rewards can be fathomed.[6] Simply put, fairness judgment requires some ability in quantitative cognition.

This requirement poses little barrier for a wide variety of animals, ranging from insects to mammals, including some that are not considered "brainy." Many have a keen sense of quantity, or, in psychological lingo, mathematical cognitive ability. This is especially evident in mate choice among females able to sense nuance in the quality of potential mates. Some female butterflies swoon for males with dazzling iridescence while avoiding the less colorful. In the guppy, a tiny freshwater fish, females shun drab suitors for bright males with the most orange spots. In the Túngara frog in Panama, females favor males whose love songs are studded with low and loud chucks. Peahens prefer peacocks with long tail feathers decked with beautiful eyespots, rejecting would-be consorts with shorter tails.[7]

Birds in particular often have a clear, sometimes finicky numerical sense. This trait is likely related to, and perhaps evolved from, the mundane task of counting eggs in their nests. It is not in a bird's nature to brood in earnest until the number of eggs reaches a certain point. Several species of eagles, for instance, lay only one egg; pigeons, two; robins, three to four; starlings, five; mallards, typically seven to ten. This is no secret to avid birders, who at times use clutch size—the number of eggs when the nest is full—as a species-specific character. Many birds, such as parrots, ducks, and chickens, can be tricked into laying more eggs if some are removed from their nests before the characteristic clutch size is reached.[8] It appears that these birds keep counting and, within their reproductive limitations, adding more eggs to make up for the mysteriously disappearing ones. This mental loophole can be exploited by breeders, who incubate extra eggs by artificial means.

The accuracy of a bird's numerical sense depends on its natural history. In the American coot, hens often stealthily lay eggs in others' nests—a phenomenon called brood parasitism, seen in well over a hundred species of birds—in the hope that they can fool the unwary hosts to work as free nannies. These parasitic hens can thus raise extra chicks without assuming the laborious motherly duty themselves. Under the pressure of widespread brood parasitism, natural selection favors coot hens that have keen numerical senses and greater visual sharpness to recognize spot patterns on eggs. Equipped with these cognitive acuities, they can, with impressive accuracy, count and identify their own eggs while rejecting those snuck in by others.[9]

Mammals are even more closely scrutinized by scientists for their numerical sense. Behavioral studies have revealed that rats and monkeys have varying levels of quantitative ability. Many have at least some basic numerical sense, often represented as mental magnitudes. This means the animals can comprehend, if not precisely, what is more and what is less. Several species of monkeys even show rudimentary skills in simple arithmetic such as addition, subtraction, and putting numbers in order.[10] This numerical sense is often relevant to survival. For example, chimps cue on pant-hoot vocalization to assess the size of an enemy troop. Only when they are sure they outnumber their opponents will they launch an attack. It seems that mammalian brains are far more sophisticated than what is minimally required to perform simple quantitative comparisons.

The neural processes of quantitative cognition and comparison are complex. Fortunately, they are among the best understood, involving special kinds of neurons in the midbrain that, when stimulated, release the neurotransmitter dopamine. These dopamine neurons are responsive to information collected by the brain that is related to rewards.[11] A closer look shows that these neurons come in two types. One decides how to act. It gets excited by rewards and is inhibited by punishments. The other type makes no distinction between rewards and punishments; this type responds only to the size of the stimulus—large reactions for large stimuli, small reactions for small stimuli—and so decides whether and how intensely to act.[12] Putting the scenario in the context of Brosnan and de Waal's capuchins, the first type of dopamine neuron prompts the monkeys to either accept or reject the rewards by comparing what their peers get—equal or better, accept; worse, reject. The second type determines how strongly they should react (accept or *gladly* accept; reject or *angrily* reject, for instance) according to what they expect and what they actually get.

This minimalist view of the two types of dopamine neurons is inadequate for a sense of fairness, however. Otherwise, many animals, especially birds and mammals, would act as if they all knew the fairness principle of "equal pay for equal work" in its primordial version—simply because they have similar dopamine neurons. If the experiment with capuchin monkeys tells us anything about their sense of fairness, it is in the tantrums they throw when shortchanged. Here, emotions are critical in signaling what they perceive as fair or unfair. But emotional signals like tantrums have to be understood by peers in context—for example, the context of carving up a hunting spoil. Otherwise, the barrage of negative sentiments unleashed by the feeling of injustice would be wasted. Worse still, since negative emotions—such as envy, anger, resentment, and hatred—are harmful to the body, they would stand little chance of evolutionary success if there were no benefit to offset the cost. When offended, we humans might experience a rush of hormones and a surge in blood pressure while we twist our face into a snarl. All this would be worthless, or even confusing, if people around us could not make out why—that is, the cause and context of our display.[13] So, too, in the animal kingdom; an animal with a sense of fairness employs its emotions to get its message across to its peers. To do so requires a crucial cognitive component that transcends the ability to count or measure—empathy, the ability to perceive the feelings of its peers. In other words, to have a sense of fairness, an animal must be capable of experiencing the emotions of its social partners.

How does the brain experience the feelings of a peer? In the vast jungle of the 100 billion neural cells in the human brain, scientists have tracked down a special type, called mirror neurons, that underlie the ability to experience empathy. These neurons occur in many areas of the human brain and are activated when a person sees, and is triggered to imitate, the behavior of another. Mirror neurons, as the name indicates, allow us to feel other people's emotions by mentally swapping our perspectives with the perspectives of those around us. These "mind-reading" cells seem to be crucial for empathy because they "allow us to grasp the minds of others not through conceptual reasoning," explains neurobiologist Giacomo Rizzolatti, "but through direct simulation," that is, "by feeling, not by thinking."[14]

Mirror neurons were found in monkey brains—the premotor and parietal cortices in particular—as early as the 1990s. Like humans, monkeys respond to the emotions and assess the intentions of their social partners.[15] But there is another layer of biological intricacy underlying empathy in humans and apes. In addition to mirror neurons, scientists have discovered another type of neurons, spindle cells, in the prefrontal cortex, just behind our forehead. Not surprisingly, the brain region is directly wired to, and presumably receives information from, the dopamine neurons in the midbrain.[16] Spindle cells also appear critical for empathy. In people with autism, spindle cells are found in a different location in the brain. Consequently, the lack of empathy—the most debilitating mental deficiency among autistic people—may be related to abnormal neural wiring, leading to difficulty with social interactions. Interestingly, spindle cells are also present in several species of whales, indicating that these intelligent animals may also have the capacity for empathy.[17]

Although optimism abounds regarding our understanding of the neurobiological foundation of empathy, there is much to learn about the roles of mirror neurons and spindle cells, and their links with empathy. Although we have yet to know what minimal level of complexity in neural networks is necessary for the emergence of a sense of fairness, existing evidence points to fairness being an evolved behavioral trait that connects humans with other species, especially primates.

If empathy itself is bound up with social life, how important is social living to the evolution of fairness? Before answering this question, let us ask a more basic one: Why are some animals social while others are solitary? A common approach to this question uses a cost-benefit analysis borrowed from economists. The costs of social living include a heightened level of competition for resources—food, shelter, water, mates—and a higher risk of contracting germs and parasites from peers. The benefits, however, can also be sizeable. They include coordination and collaboration in searching and hunting for prey (as in wolves and lions), sharing food (as in many monkeys and apes), defending against predators (as in meerkat sentinels), protecting territories, and mutual cleaning and comforting, to name a few.

Cost-benefit analysis is quite useful for understanding evolution. Natural selection mandates that, all things considered, if the benefits of social living are less than the costs, animals should live alone. But clearly, many animals stick together in groups; why, if they do not profit from the company of others? To be social implies that, by living with peers, they gain more than they would by living alone.[18] Observation of social animals demonstrates that the benefits of social living—coordinated hunting, food sharing, joint defense, or mutual protection—pivot on cooperation among partners. Cooperation is, therefore, the main drive of social living. Without it, how can it be possible to have an organized society with stable members?

The necessity of cooperation for social living is best illustrated by what I call pseudo-social animals—gazelles, wildebeests, zebras, buffalos—on the savannas of Africa. These animals rarely cooperate with each other; stable social bonds seldom occur beyond mother-young relationships. Attracted to food or water, their congregations can be impressive, but they will disband and disperse in every direction once their hunger and thirst are sated or they are chased by hungry predators. This is not the kind of society that favors cooperation. Without cooperation, the benefits of social living cannot outpace the costs; society dissolves.

Social living brings conflict of interest between peers to the fore, however. If not properly resolved, conflict of interest can be the downfall of any society. It can derail cooperation and dissolve hard-won relationships, as many experience in marriages, friendships, associations, and joint business ventures. For this reason, conflict of interest may be the mother of all moral problems and the hub of the biological quest for morality that Alexander argued for. To enjoy the "sweet spot" of cooperation, many social animals, humans included, have evolved a battery of strategies for conflict resolution. When physical disputes occur, animals will often go to great lengths to repair wounded relationships. Wolves in a pack lick each other for reconciliation, and monkeys and apes engage in mutual grooming—much more than usual—after an outbreak of fights. Although we are not entirely sure about other primates, humans, at least, are equipped with a special emotional tool—feelings of guilt—to help keep relationships in working order. Often displayed after violating moral rules in general, and particularly when treating peers unfairly, guilt motivates us to restore peace and harmony with our social partners. These are just a few examples indicating the importance of affable relationships for lasting cooperation.

Cooperation is propelled by two biological engines: kin selection and reciprocity. Kin selection—a theory developed by biologist William Hamilton in 1964—explains how altruistic behavior toward genetic relatives, which at first glance may seem costly for the altruist, will eventually be a benefit.[19] Put simply, altruists help themselves—that is, they increase the chance that genes they carry will be passed on to the next generation—by helping their genetic kin. This can be illustrated using a hypothetical example. Say I have a sister, who on average shares 50 percent of her genes with me. Kin selection, as a form of natural selection, shows that it makes no difference whether I help to pass on the genes in myself or those in my sister. As long as more copies of the same genes make it into the next generation, they stand a better chance of thriving in the future. Thus, as long as the cost of my sacrifice—any sacrifice, from lending a hand to laying down my life—exceeds twice the benefit for my sister (as she carries a half of my genes), it is a better deal for me to make the sacrifice than to give my sister the cold shoulder. To make me give up my life, I need to save more than two siblings to gain a genetic advantage for my sacrifice. All in all, my sacrifice is ultimately self-serving.

According to Hamilton's kin selection theory, the more genes blood kin share, the more likely they are to stick together and cooperate. This explains why humans tend to be much more dedicated to our close relatives, such as children, brothers, and sisters, than distant ones, such as cousins and nephews.

In ancient China, kin were so meticulously sorted out that relatives as distant as five or six generations up and down the lineage tree were identified in large clans, often with hundreds of clansmen living together. For any man, the imperial government legally recognized six degrees of his kinship—more substantial than ever existed in the West—based largely on genetic closeness.[20] One major motive behind this system was the inheritance of family assets—land, houses, money—after the death of a clan patriarch. Not surprisingly, valuable properties were carefully divided and distributed in proportion to the degree of kinship. By sticking together, a clan could outnumber and outdo rival clans in scrambling for resources and providing security.[21] This Chinese example illustrates how kin selection works and why genetic relationships are important in human social living.

While kin selection is an elegant explanation of how self-sacrifice works between those who share their genetic makeup, altruistic behavior is also common between genetic strangers in many societies, where kin and non-kin are mixed. Without the binding force of kin selection, how is this possible? Evolutionary biologist Robert Trivers developed a theory to account for this paradox in 1971. He proposed that, besides kin selection, another pillar for cooperation is reciprocity, which is exemplified in the common saying, "You scratch my back, and I'll scratch yours." Trivers's reasoning is straightforward and elegant: if two parties can each gain more by working together than by going it alone, cooperation will prevail.[22] Indeed, reciprocity among social partners is common in primates, ranging from literal back-scratching to more sophisticated coalitions and alliances. Since reciprocity can occur regardless of genetic relationship, it is therefore more ubiquitous than kin selection in promoting cooperation. By extension, members of a group of kin, held together by the dual glues of kin selection and reciprocity, are more prone to cooperate and more eager to come to each other's aid than members of a group without blood ties. Here, the common saying "blood is thicker than water" is still valid.

Kin selection, though often a more potent force for cooperation than reciprocity, only works for a small circle of blood relatives. It is obviously ineffective as a means for promoting cooperation in groups where most of the members have no blood ties. Reciprocity, on the other hand, shows no such constraint and is often the main driver for large-scale cooperation among humans, for example, in teamwork, charity, patriotism, and international alliance. Accordingly, reciprocity should be given its due in our search for the connection between social living and fairness.

Before moving on, let's first return to the question of why reciprocity is normally a weaker force than kin selection in promoting cooperation. Far from being foolproof, reciprocity has a major loophole—it is extremely vulnerable to cheating. Reciprocity is supposed to be a mutual process, swapping favors between partners. But the time lapse between donating and receiving a favor opens an ominous window for recipients to shirk their duty and fail to reciprocate. If a favor is not returned, the original donor suffers a net loss.

Clearly, one has to be wary when picking partners. Cooperation without assessing the likelihood of reciprocation is, in ecologist Garrett Hardin's words, "promiscuous altruism," and makes the cooperator an easy mark, likely to be preyed on by cheaters. For this reason, even Pierre-Joseph Proudhon, the father of anarchism, who pegged his faith to kindness in human nature, universal brotherhood, and eradication of government, approached altruism with guarded precaution: "If everyone is my brother, I have no brothers."[23]

Cheating is like a blood-sucking parasite that saps the vitality of cooperation. Reciprocity, to stand any chance of survival, needs some potent antidotes. These antidotes, in Trivers's prescription, include long life spans and stable group memberships. A long life span can provide ample opportunities for social animals to interact and discover the sweet spot of reciprocity; stable group membership helps build trust and deter cheating. In the best scenario, everyone knows everyone else within the group. Cheaters, like the fabled boy who cried wolf, cannot fool others for long after everybody in a community becomes aware of their duplicity. In other words, face-to-face interactions inhibit cheating. Not surprisingly, people tend to be more trustful of one another and behave more honestly in villages and towns than in large cities. Small communities, albeit often lacking in diversity and dynamism, typically enjoy lower crime rates and less frequent cheating than large metropolises.

Humans are not alone in these considerations. Vampire bats, contrary to their ghastly name, show ample reciprocity within trusted circles. After years of observation, biologist Gerald Wilkinson found that if a bat does not suck blood for seventy-two hours after a big meal, or after just three nights in a row of failed hunting, it will starve to death. Although vampire bats work hard for their nightly meal, a successful hunt for blood is far from certain. In fact, a third of bats younger than two years old will fail on any given night. To prevent starvation by spreading the risk, cavemates that roost in close vicinity operate a food-sharing system. The successful and satiated bats regurgitate blood to the famished. Bats in a trusted circle mostly know one another, in addition to being, to some degree, genetically related. This system protects the cooperating bats from cheaters, also known as free riders—those who might come to share blood, but either do not return the favor at all, or return it less often than they should.[24]

In vampire bat communities, cheaters, if detected, may suffer a dire consequence—death by starvation. Thus, cheating is easily deterred by simple rejections of food-sharing requests. In most situations, however, stronger measures are needed to reinforce the trust between cooperating partners. Just as penalties—such as higher interest rates and potential legal actions—are vital for the smooth running of our credit system, some basic behavioral rules are essential to ensure harmony in animal societies. For cooperation to work, transgressions against these rules must incur penalties in the form of costs to the transgressors. Here, the cliché "necessity is the mother of invention" is a good description for the emergence of moral behavior in animal societies. The need is cooperation; the inventor, mutation followed by natural selection.

Fairness, in particular, can arise as a set of behavioral rules to resolve conflicts of interest, and it evolves through penalties when these rules are breached. Violators can bear a variety of unpleasant consequences, including losing the benefits of cooperation and likely suffering physical retaliation. The victims of unfairness, accordingly, have two ways to fend for themselves: leave the group or fight back. Both will exact costs from the violators. However, many highly social animals have evolved to the stage where they can no longer survive alone. For them, leaving the group is rarely a possibility. Fighting back—the only option left for the injured party—offers a range of punishment choices, from expressing nuanced displeasure or dissolving alliances to physical retaliation such as chasing and biting, commonly seen in primates. More often than not, however, violators will fall back into line with merely a cross facial expression or a grumble from the victims, allaying the need for further, more consequential actions. A simple display of emotions resulting from a perceived injury is often enough to convince a violator to stick to the rules of fair play.

As it takes at least two for fairness to matter, solitary animals, even the brainy ones, have no need to develop a sense of fairness. For example, what use would a sense of fairness be for the tiger, a largely solitary species outside the breeding season? Despite its intelligence, a sense of fairness has little utility for an animal that does not live socially. Hence, social living is a necessary ingredient for a species to evolve a sense of fairness. Without social living, fairness becomes irrelevant—it confers no adaptive advantage, and so natural selection has no power to boost its evolution.

Through the lens of evolution, we've now identified an abundance of new insights into a long-standing controversy that has been raging since the days of the ancient Greek philosophers: the origin of fairness in humans. Before the discovery of fairness in nonhuman social animals, some scholars considered fairness a divine endowment; others thought it had to be acquired through moral teaching. As scientific evidence accumulates, it is increasingly clear that fairness—both in animals and humans—is an adaptive trait that evolved for settling conflicts of interest, the inevitable consequence of social living. This new perspective has its philosophical roots among a particular school of thinkers, the social contractarians—including Rousseau, Kant, and Rawls—who believe fairness is a social agreement, as it embodies a distinct feature of social contracts. Fairness often invokes tradeoffs between different preferences. Animals may give up some benefit, such as cucumber slices in the case of the capuchin monkey, in exchange for fairness. Likewise, humans are generally content with making some sacrifices in order to increase fairness and justice in society.[25] But agreement on what actually constitutes fairness is another matter.

As we have seen throughout the animal kingdom, social living occurs when the benefits outweigh the costs for all members of the group over their lifetimes. Yet a social group is made up of many individuals, each bidding for its own highest possible net gain. Such fitness maximization is mandated by natural selection—a blind force that punishes individuals for putting in even a mildly subpar effort in passing down copies of genes to the next generation. To realize the best outcome of social living, an animal has to strive for its best to gain an edge over rivals when interests conflict. Reciprocity, too, can be rife with antagonism, where either party may attempt to outsmart the other for even a blade-thin margin in any transaction. This means that he who does not fight for his own plate will suffer and over time be gradually edged out in the evolutionary race by those who do.

In such a system, cooperative partners may be torn amid the conflicting motives of gaining more now, on one hand, and losing potential partners for future cooperation on the other. In the long run, fairness in reciprocal exchanges can satisfy both parties. As a result, this happy medium has naturally evolved to be an equilibrium point. This may explain why fairness is vital for keeping trading—the central activity in the economic market—going in human societies. In fact, trading, in essence, is a form of reciprocity by which both parties can get what they want, often at the same time.

The reciprocal nature of trading is even more obvious in bartering without the mediation of money. Goods and services are directly exchanged, reminiscent of the reciprocity observed in many social animals. Biologists Ronald Noë and Peter Hammerstein believe that reciprocal exchanges seen in many animals, especially primates, are indeed trading, governed by the forces of supply and demand in the biological market.[26] This idea has motivated primatologists to reexamine reciprocal exchanges, especially in social grooming, by which animals take turns scratching each other. In addition to the obvious benefit of ridding them of parasites and soothing itches, grooming in the primate world can lower stress hormone levels in the body and boost endorphins, a family of happiness hormones, in the brain. Thus, the need for grooming, especially in hard-to-reach areas of the body, is a service in high demand that can be filled by a social partner. With such a strong and broad demand, an animal that provides the service can use grooming as an all-purpose currency to barter for a variety of other goods: obtaining consolations, being scratched and comforted, mating opportunities, gaining permission to hold infants, reconciling wounded relationships, building or consolidating alliances, and currying favors from higher-ups.

Although studies that use the biological market perspective have only recently gained momentum, there is no shortage of excellent examples. In both chacma baboons in South Africa and Tibetan macaques in China, females of the same ranks trade grooming in close to even amounts: more for more and less for less. But when such trading occurs between females of unequal social status, the larger the gap between their ranks, the lower the reciprocation from the higher-up. Presumably, the lower-ranking female hopes for something else beyond grooming, such as favorable treatment as compensation for the extra amount of scratching given to the higher-up.[27] In the long-tailed macaque in Indonesia, males, when their minds are set at mating, will provide longer grooming service to females as a premium in exchange for sex.[28] This is also the case in a less brainy social lemur, the sifaka, in Madagascar. Before the mating season, males only groom themselves. But when the mating season sets in, males begin to barter grooming for sex with females. This seasonal shift in the grooming pattern of males is telling, reflecting the oscillation in the price of sex over time.[29] Here, social grooming bears an essential sign of trade: the fluctuation of supply and demand, a distinctive feature of markets.

If grooming can be traded for so many different goods among primates, we can expect, in spite of limited data at this point, that other commodities such as food, mutual aid, coalitions, alliances, sex, and other resources can be bartered as well. Even more so than grooming, sex is a nearly universal commodity that can be exchanged for resources by females in a variety of species. Primates aside, trading sex for resources—sometimes euphemized as nuptial gifts—is seen in animals as simple as flies and crickets. In the hangingfly, a female will accept a courting male that offers a blowfly as a nutritious nuptial gift. But the female "charges" by the minute. The time the female allows the male to mate with her depends on the size of the blowfly; the larger the blowfly, the longer the copulation, quid pro quo.[30] This textbook example shows that sex can be a highly valuable commodity in the biological market. For this reason, prostitution is often jokingly referred to as the oldest profession, for its origin in the biological market may be traced back hundreds of millions of years to a time before the dawn of humanity.

Trading is cooperation writ large, though we are losing this sense in the age of e-commerce, when face-to-face bargaining is on its way out of our consciousness. Extensive trading would be unthinkable if the exchanges were not perceived as fair by market participants. The reason for fairness in trading is simple: trade between willing parties results in a win-win outcome. Since each party intends to exchange what the other needs, it's better for each to make the trade happen than not—unless the bid or offer is unacceptable for one or both parties. Since consistently unfair trades will eventually put one out of business, trading, as a form of cooperation, is unlikely to evolve without fairness.[31]

In Brosnan and de Waal's study, capuchin monkeys do not have to know the face value of a token used for trading—though with patient coaching they can learn some human-like tricks of trade for money or sex.[32] In the experiment, they simply need to make the comparison between what they get and what their social partners get for the same trade, because the "price" of their token money is set by the "market value" at the moment. We can safely assume that the monkeys would do the same if diamonds, instead of stones, were used as tokens. It's not difficult to see that trading in the biological market is a major force behind the evolution of fairness in social animals, whose brains are sophisticated enough to have some basic capacity of quantitative cognition (for judging what is fair and what is not) and a certain level of empathy (for understanding the emotions of peers).

There is another twist in the link between social living and fairness: social living is not all that matters for the evolution of fairness. We know that a social life is also a hierarchical life. No society in the animal kingdom is completely equal for all members. Among humans, high or low status is a significant factor in even the most egalitarian societies. In modern democratic nations, one can hardly miss rank order in most social settings—schools, governments, corporations, and even casual gatherings. Hierarchy, in essence, represents inequality by implicit agreement or, in the case of humans, often by social institutions. Higher ranking individuals typically reap more benefit than those on lower rungs. In governments, private companies, military units, or public schools, leaders—presidents, governors, chancellors, generals, CEOs, or principals—almost always earn more than ordinary members or employees, sometimes by hundreds of times. Likewise, studies of wolves, monkeys, and apes consistently produce results showing that higher ranking individuals have better access to food, water, shelter, and mates, which translates to a higher level of fitness than that of lower ranking members of the group.

What should the rank and file do with the situation? On one hand, it behooves them to reject the ordained fate of being taken advantage of in life. Natural selection mandates them to fight back—one way or another, one time or another—in order to advance their own status. If they don't, the genes they carry may face elimination from the gene pool over time. Behavioral studies in a large number of social animals, particularly primates, show that lower ranking individuals do sneak feeding and mating opportunities that violate the interests of higher ranking members in the group. And, when the time is right, they challenge or revolt against higher-ups. Conflicts such as these send the social hierarchy into a state of flux. A wobbly or chaotic hierarchy typically hurts higher status individuals more than lower status ones because the former have more to lose than the latter. In lions and langurs, for instance, dethroned alpha males lose their harems and their young offspring to usurpers. In traditional human societies, when monarchies and dynasties were overthrown, kings and emperors were often killed, together with their children, relatives, and, if not taken by the victors as reproductive trophies, wives and concubines as well.

On the other hand, even though conflict of interest is a recurring factor of social life, constant strife heightens the cost for most, if not all, members, and it can destabilize the group. An unstable hierarchy, one rife with such conflicts, involves a high cost for social animals, as is often seen during leadership turnover in primates. In humans, an unstable or transitional period in organizations—schools, companies, local governments—will likely be accompanied by uncertainty, anxiety, and even chaos. At the national level, in extreme cases, it may lead to mass violence. In recent decades, we have seen such calamities in Latin America, the former Yugoslavia, Indonesia, Rwanda, Egypt, Libya, Syria, and many other countries and regions, not to mention the scores of bloody revolutions throughout human history. A stable hierarchy, even though it calls up the distasteful feeling of inequality, is not all that bad from a practical point of view for the majority of the members of the group. Peace and stability are usually in the collective interest of most members, enabling them to conduct their basic biological business—that of realizing their potential fitness.

In the end, a social hierarchy implies a truce between the two opposing selective forces in resolving the perpetual quandary of conflicting interests. Each and every member in the group wants to be the king of the mountain, yet each has to make necessary concessions to others in order for peace to prevail. To arrive at a compromise under varying social and ecological conditions, different species, different populations of the same species, or even different groups of the same population may take vastly different routes. Consequently, nature exhibits a large spectrum of hierarchical patterns, from complete dictatorships to highly egalitarian societies.

In our quest for understanding the evolution of fairness, the piece still missing from the jigsaw puzzle is an answer to the question of why fairness is selected for within the context of social hierarchy. To answer this, we can imagine ourselves as males in a gorilla society. Gorillas live in close-knit, hierarchical groups with silverbacks as the leaders and all others on the rungs of the ladder. In our gorilla world, we have no concept of Darwinian selection. Our ingrained instinct urges us to be the king of the mountain, despite our ignorance of the theory. With a distinct advantage—access to food, water, shelter, mates—over all others in the group, power is seductive. But the reality is a cold one: among all twenty members in our group, there is only one alpha male, the silverback. For the remaining nineteen of us, all but one—the wretched omega—are both victims and victimizers, being exploited by those who sit above and, at the same time, exploiting those who sit below. While few of us would complain of having too much power, which comes with high social status, most would grumble when being taken advantage of by the higher-ups. Therefore, except the alpha male, everybody else in our gorilla society should fight for a higher status. Even the alpha male shares our feelings before he is crowned the king of the mountain. He will surely fight against new kings after he is dethroned down the line.[33] This "gorilla confession" about their secret social lives illustrates why the sense of fairness can evolve in a hierarchical society. Fairness here constitutes a bottom-up leveling force, pushing toward equality in resource distribution in the hierarchy. We can expect that the steeper and more rigid the hierarchy is, the stronger the selection for fairness becomes.[34]

What works for gorillas is, by and large, what also works for humans. Our perception of fairness tends to vary with our social status. Business leaders often prefer large pay disparities and low taxes; blue-collar workers, meanwhile, are inclined to support more equal pay scales and taxing the rich. On the issue of allocating pay raises, people, when assuming the role of a supervisor, tend to emphasize fairness in the procedure, so that they have an advantage in the outcome over the rank and file. (That's why CEOs often claim how important they are in the running of their companies.) But when put in the shoes of a subordinate, they care more about fairness in the outcome, that is, even division of resources.[35] (In fact, fairness in procedure—procedural justice—and fairness in outcome—distributive justice—involve different brain regions, a discovery we will discuss in the next chapter.) Likewise, in families where husbands contribute more to family incomes than their wives, the husbands are apt to stress their priority in access to money for personal spending whereas the wives place their emphasis on equal sharing of family financial resources.[36] Even in the global arena among nations, the situation is similar. Winston Churchill, when reflecting on the link between national strength and the international pecking order (which still holds true today), was brutally honest: "The whole history of the world is summed up in the fact that, when nations are strong, they are not always just, and when they wish to be just, they are no longer strong."[37] Here one can hardly miss the ubiquitous bottom-up push—selection—for equality in social hierarchies. Even for higher-ups, the very prospect that they, too, can fall to the lower rungs of the hierarchy one day makes them prefer a certain level of fairness.[38] After all, what goes around comes around.[39]

Prodded by commonsense questions, we have traveled a considerable distance in tracing the evolutionary genesis of fairness. By now I hope I have presented enough evidence to support this conclusion: *fairness, as a mental instinct and behavioral rule for solving conflicts of interest, is spawned from social living and social hierarchy. Behind it are two major selective forces, reciprocity for mutual benefit and compromise for social harmony, both of which are critical for maximizing the net benefit of cooperation.*

Despite the dazzling creativity of evolution, however, not every trait can evolve. It would be nice if we could produce purple-haired descendants by dyeing our hair or endow future generations with wings by skydiving and bungee jumping. However, natural selection can only work on a trait that is heritable. Hence, fairness, if it is indeed subject to natural selection, must show its genetic implications. If evidence does demonstrate that fairness is linked to genes, it will bring up two crucial questions. How is the fairness instinct mapped onto our behavior? How then does fairness-motivated behavior affect our society? To answer these questions, we turn to the next chapter.

Chapter 3

ROBIN HOOD IN OUR DNA

Driven by our fairness instinct, we keep watchful eyes on all sorts of violations of fairness rules in our daily lives. A person who cuts in line raises eyebrows or invites a rebuke. Even more illustrative are driving disputes. We can instantly be ticked off if a car that arrives later, even by a couple of seconds, moves first at a four-way stop. Minor traffic violations may ignite road rage or incite tailgating by the offended party. Likewise, we can anticipate reactions for failing to reciprocate on a party invitation from a colleague, for forgetting a significant other's birthday, when a coworker receives a raise in salary when they have done nothing to deserve it, or when an inept crony of a supervisor is promoted. These, together with myriad other examples, rouse resentment and indignation, often with a minimal amount of rational thought involved.

Our subconscious emotional response in these common scenarios indicates that a preference for fairness may be deeply ingrained in our DNA. This hypothesis has gained much support from scientific studies in recent years. One area of evidence comes from game theory—specifically, an exercise known as the Ultimatum Game, designed to test people's sense of fairness. In this game, two people, call them Peter and Richard,[1] are given a certain amount of money, say $10, to divide between them.[2] Peter makes a proposal for the division between himself and Richard—for instance, 9:1, 7:3, or 5:5. If Richard agrees to this division, the deal is done, and each keeps the money as proposed. If Richard rejects the proposed split, the deal is off, and neither of them gets anything.[3]

In experiments with real people, the game is typically set so that Peter and Richard have never met before, nor will they ever meet again.[4] This is a one-time deal between two strangers. Under these contingencies, how much will Peter offer to Richard? From Peter's point of view, the more he keeps for himself, the more he gains; yet a greedier proposal makes it more likely that Richard will reject his offer. Thus, the more he proposes for himself, the greater his risk of getting nothing. Richard also faces a dilemma: how stingy an offer should he accept? Even without considering the Hobbesian worldview that human nature is essentially selfish, Richard, led by his intuition, cannot expect that Peter would give him more than $5. (This "impossibility" can happen under some specific circumstances, as we will see later.) But will he accept $1, $2, $3, or $4? A once-popular economic idea called "rational choice theory" tells us that, for Richard, even one dollar is better than nothing. As a thinking person, he should accept any nonzero offer from Peter. Is this the case in reality?

The results from industrial societies say no. For those who play Peter, the offer is typically above 30 percent, or a $7:$3 split, with the mean between 30 and 40 percent. The most common offers are 40–50 percent, or nearly even splits. Richard, on average, rejects the proposal about half the time if Peter's offer falls below 20 percent.[5] Here, we can see that those who play Peter are, surprisingly, more generous than expected, considering they can offer a lot less to those playing the role of Richard. How does this real-world Peter think when offering Richard such a generous proportion? Is he afraid that he may lose everything because Richard has a trump card in his hand—rejecting the proposal? Phrasing the question differently, is Peter prone to play it safe? If we take away the trump card from Richard, what then would Peter do?

To answer these questions, let us make a small modification to the rules: Peter can still propose whatever money split he prefers, but Richard is stripped of his voting power. Richard now has no other option but to take Peter's offer. With Peter the de facto dictator, the Ultimatum Game becomes, fittingly, the Dictator Game. He now has a monopoly on power. Will Peter claim everything for himself?

The answer is—even more surprisingly—no. Studies show that when given the option of either a 9:1 or a 5:5 division, three-quarters of people who play Peter opted for the even split.[6] Even in a study in which people appeared stingy, they still offered Richard 20 percent, even though they could have kept everything for themselves.[7] Clearly, most people in Peter's shoes do not consider themselves callous dictators, despite the power at their disposal. Why is their behavior at odds with the behavior projected by rational choice theory, in which the Peters, after conscious and careful reasoning, should hog all the profit and leave the Richards with zilch? Why do people deviate from so-called rational choice, foregoing potential profit to benefit a stranger they will never again meet?

The disparity between reality and rational choice theory appears to stem from the fact that both experiments, the Ultimatum Game and the Dictator Game, are designated as one-shot situations. Peter is supposed to take all on the table and disappear without trace. However, though the researchers go to great lengths to emphasize that Peter will never meet Richard again, our brains have evolved under conditions in which such anonymity was not normally the case. Despite the dazzling pace of cultural change, much of our instinctual behavior is still tuned to life in the Stone Age, when our ancestors were living in small, close-knit bands—not unlike those of today's chimps. In this ancestral environment, if you encountered another individual, it was unlikely that you would never meet him again—as the game theory researchers presumed in their studies. This point is easy to get by asking a simple question: how far do people normally travel by foot? If the answer is not much, then how likely is it that two people living within walking distance never meet again?

How might a mentality tuned to such long-term vision be favored by natural selection? One obvious answer is that Peter's largesse can make Richard feel obligated to return the favor in the future. Reciprocation reinforces their relationship; both Peter and Richard gain benefits over time. A less obvious factor favoring a long-term perspective is reputation. Peter's reputation equates to his trustworthiness, the critical distinction between cooperator and cheater. Reputation thus serves as a badge of how cooperative one is. Evolutionary mathematicians Martin Nowak and Karl Sigmund show that a person with a good reputation often gets help from people for free—free in terms of not having to return the favor.[8] Therefore, in a stable community, a good reputation is, in a sense, a money tree. The better its cultivation, the more can be harvested over time. In other words, a good reputation earns long-term profit.

The intrinsic value of reputation can explain many seemingly enigmatic human customs, from the tradition of dueling in Europe, to "saving face" in East and South Asia, to honor killing in some Muslim communities. For the same reason, the ancient Chinese practiced an elaborate culture of filial piety, where family elders were revered to a degree that they were entitled to the best living conditions—the finest food, the coziest accommodations, and the highest social status—even though they were too old to work.[9] The logic: how can we trust a person who won't even treat his parents or grandparents well? Filial piety served as the gold standard for reputation in traditional Chinese communities. Without an understanding of the origins and value of reputation, the practice of filial piety makes no sense.[10] With it, we can see how the benefit of reputation warrants the cost of sacrifice.

Today, even without conscious learning, our genes, through their expression as our instincts and intuitions, still send the old, stubborn instructions to our minds, making us behave as if we, the Peters, would encounter the Richards again in the future. This subconscious—and otherwise irrational—perspective coaxes us to defy the temptation to burn bridges when there is a good chance we'll need to cross them in the future. Hence, no matter what efforts the researchers make to convince the subjects they will never meet again, those who play Peter seem stuck in a mind-set opposed to theoretical anticipation. Feeling quite at odds with this new, one-shot game, their brains tend to slip back to ancestral modus operandi, whispering, "This is an opportunity to build a new friendship for future cooperation. Don't lose it." In fact, our natural penchant for cooperation is so ingrained that when we share, donate, and act fairly the brain areas associated with reward and positive reinforcement learning are mobilized; we feel happy and satisfied as a result of our actions.[11] No wonder people in Peter's shoes tend to be more generous than rational choice theory would suggest.

Our brains motivate us to behave differently than rational choice theory predicts because evolution has equipped us with a subconscious long-term perspective that overrides our conscious short-term reasoning.[12] This may explain why even children demonstrate a high level of fairness before they are fully integrated into the adult world, as shown in the following study in public schools in Oregon:[13]

Table 1. Rates of offer and rejection in the Ultimatum and Dictator Games

Age (Year)	Grade	Ultimatum Game Offer (%)	Ultimatum Game Reject (%)	Dictator Game Offer (%)
7	2nd	35	11	3.5
9–10	4th	41	14	14
14	9th	45	3.3	12
17	12th	43	10	21

Modified from data presented in W. T. Harbaugh, K. Krause, and S. G. Liday Jr., "Bargaining by Children," University of Oregon Economics Working Paper No. 2002-4 (March 1, 2003), p. 11.

In Switzerland, when children were allowed to choose between an even and an uneven split of their favorite European sweets (Smarties, Jelly Babies, or Fizzers, which, for young kids, have higher value than money), 21 percent of the 3–4 year olds, 33 percent of the 5–6 year olds, and 60 percent of the 7–8 year olds preferred the even split.[14] The smaller offers by younger children in both studies are particularly telling. They are signatures of an innate gut instinct before reason kicks into full gear, as children grow up in the industrial world.[15] Here, the proverb, "the child is the father of the man," is almost literally true. Despite the less generous offers in young children, the willingness to share, regardless of cultural influence, signals a tendency toward egalitarianism entrenched in our biological nature.

The subtle yet fatal flaw in rational choice theory is that it overlooks—and thus discounts—the influence of our subconscious motivation for fairness. Our pride in our rational minds can make us ignore our hidden instincts, which may better reflect our evolutionary past. Evolutionary biologist Steve Johns coined a metaphor for these ancestral instincts that still strongly influence our behavior today, calling them Darwin's Ghost. The counterintuitive results revealed by the Ultimatum and Dictator Games show that reason is only one influence on our behavior, and perhaps not the strongest one. Our brains are still tuned to long-term human relationships in prehistoric societies. When the historical context of human evolution is taken into consideration, the word "rational" in rational choice theory is, in reality, misguided.

Why do so many people who play Richard reject a 20 or even 30 percent offer? In addition, a recent study reveals that there is a genetic component—over 40 percent as measured in heritability—in the tendency to reject unfair offers in the Ultimatum Game.[16] How could evolution favor a genetically based instinct to reject a sizable benefit? Wouldn't even a minuscule gain that can be used to promote fitness—measured by the number of surviving offspring—be better than nothing in an absolute sense? Why wouldn't natural selection eliminate this instinct?

Again, reputation may be at work. When Richard accepts a low offer—though better than nothing—his reputation suffers. Knowing his readiness to settle for less, others in the community may take advantage of Richard in future transactions. As a result, in the long run, accepting low offers turns out to be costly.[17] Rejecting low offers, on the contrary, is a worthwhile short-term sacrifice that serves to prop up a long-term public image of expecting a fair deal, or no deal at all. People will take notice and resist the temptation to shortchange Richard in the future. Ancient Chinese wisdom reflects this evolutionary imperative: never lower your dignity to accept free food, even though you are hungry. Our brains discourage us, too, from taking low offers by activating the anterior insula, an area related to such negative experience and emotions as pain, anger, and disgust.[18] Interestingly, when the right side of the dorsolateral prefrontal cortex is disrupted by magnetic stimulation, people's resistance to low offers is disarmed, even though they are clearly aware that the offers are very unfair.[19] It seems defending one's dignity is a behavior strategy as well as a state of mind.

Rejecting unfair offers to preserve his reputation can help Richard survive, enhancing his fitness, especially in a face-to-face community, where people recognize each other and interact on a regular basis. Outside the community, reputation loses its value, and taking low offers from strangers can be advantageous—in an absolute sense; it is better than nothing. So community matters for reputation, in particular, and for fairness in general. This fact speaks for an overarching yet often misunderstood or unappreciated aspect of evolution: *natural selection is more often a local than a global force*. Fitness, as the gauge of selection, is more properly calculated by comparison with peers in the immediate vicinity, not those far away. In other words, what guides evolution is *relative* fitness rather than absolute fitness.[20]

To illustrate the importance of relative fitness, assume there are equal numbers of two types of monkeys—Type A and Type B—living together. They differ in only one aspect. In its lifetime, each Type A monkey successfully reproduces and raises three offspring (fitness = 3) on average, but each Type B monkey produces four offspring (fitness = 4). After three generations, Type B monkeys outnumber Type A monkeys by a ratio of 64:27; in other words, they are more than twice as numerous. After twenty-four generations, Type B monkeys are almost one thousand times more numerous than Type A monkeys. With such a small proportion at this time, Type A monkeys are much more vulnerable to elimination, although they appear to be more numerous than before in absolute number. If mortality, due to, say, resource limitation, is the same for both types—that is, both types have equal chances of dying out—Type A monkeys will much more likely be the losers in the evolutionary race.

This hypothetical example played out in the real world on the small island of Daphne Major in the Galapagos Islands. In the 1970s, Peter and Rosemary Grant, an evolutionary biologist couple at Princeton University, noticed that the beak size of the median ground finch, one of the Darwin's finches, fluctuated over time. In dry years, all finches suffered a high mortality rate, appearing scrawny and weak from food shortages. Even so, finches with large beaks died off more slowly than those with small beaks because they could better crush large, hard seeds, the main food source available on the island. In the end, large-beak finches dominated the much-reduced population. The tables turned in wet years, however, when small seeds became abundant. Amid a booming population with abundant food, small-beak finch numbers increased faster than, and quickly dwarfed, those of their large-beak comrades because small beaks were better at handling tender seeds and more efficient in extracting nutrients.[21] No matter how the finch population waxed or waned, it was through relative numbers—relative fitness—that natural selection shaped and reshaped the beak of the median ground finch.

Fairness proceeds on a similar track, evolving under the shortsighted, local-focused parochialism of natural selection. In the Ultimatum Game, relative fitness easily explains why Richard might reject even a reasonable but uneven offer from Peter. Whatever Peter proposes, Richard's mind-set may well be tuned to relative payoff, comparing his gain with Peter's. If the discrepancy is too large, he will hesitate—even though, in an absolute sense, he would gain some from the deal. As demonstrated in the hypothetical case between two types of monkeys and the real-world example of Darwin's finches, if Richard repeatedly gives Peter too much of a competitive edge, he will eventually lose out. Surrendering a larger lead to Peter will make it harder for him to catch up. Consequently, a brutal Machiavellian tactic can emerge—*in the cutthroat competition for an evolutionary edge, ruining others' chances is just as effective as gaining an advantage for oneself.*

The calculus in relative fitness provides sound logic for why Richard becomes increasingly likely to pull the trigger of rejection as the offer moves away from being equal. In the game, Richard is prone to "getting even" with Peter.[22] (This get-even mentality—the motivation to bring down rivals' advantages—is profoundly ingrained in our sense of fairness, as I will elaborate on later.) The fairness mentality is significantly moderated when Richard plays against an inanimate Peter—a computer. A comparative study shows that when the Ultimatum Game offer from a computer was a meager $1 out of $10, over 60 percent of people happily accepted. However, when the same offer came from a person, the acceptance rate dove below 40 percent.[23] Fairness, it seems, functions mainly in regard to other people, not machines. This is another piece of evidence indicating that the fairness instinct emerged through competitive interactions among people over evolutionary time.

From a behavioral perspective, fairness requires comparison. So the importance of relative fitness cannot be overemphasized in shaping the evolution of fairness. Everyday experience teaches us that comparisons are relevant only with those we know or with those with whom we interact. This is why we more commonly compare ourselves with those around us—friends, relatives, colleagues, classmates, neighbors—than with those far away. In a related example, some Americans find the enormous salaries of professional athletes unfair, but they seem indifferent toward the salaries of team owners, who may make many times more than star athletes. This rationale is an interesting one. Popular participation in sports makes many feel they have a shot at the big time—so they can compare themselves with, say, Alex Rodriguez. It's harder to relate to the late George Steinbrenner, who made money through financial deals. It's easier to envision oneself as a highly publicized, talented player than as a wealthy owner whose lifestyle and business are too obscure to make relevant comparisons.[24]

Even though relative fitness is important to understanding how our sense of fairness evolved, we need to place it in proper context. Otherwise, we may conclude that Richard should reject *any* offer from Peter regardless of the stakes—the absolute sums of money—as long as his share falls below a certain percentage. This is not the case. Studies in Florida, Indonesia, and Slovakia show that when the stakes become higher, people are increasingly willing to take the offers that they otherwise reject.[25] Had behavioral scientists been awash with research money from, say, the Bill and Melinda Gates Foundation, they would have done a study comparing the rejection rate for $2 out of $10 versus that for $200,000 out of $1 million in an American University. It's hard to imagine the rejection rate would remain the same.

Obviously, for Richard, no matter how parochial and narrow-minded he might be, the world is much more varied and complex than his relationship with Peter; there are many other members in his own tribe alone, though they are not involved in this particular game. When the stakes go up, it becomes harder for Richard to resist the temptation to accept the offer. Otherwise, he forgoes an opportunity for a major boost in competitive advantage over his other peers. (Continue with the gorilla metaphor in the previous chapter. If the advantage is big enough, Richard could jump from, say, the sixth-ranking male to become the beta male, just below the silverback Peter.) Therefore, with high enough stakes, it's better for Richard to ignore Peter's edge for the moment. He can come back to deal with Peter later. One has to advance one step at a time. Again, it's all relative when it comes to natural selection.

In addition, the parochial nature of selection predicts that the sense of fairness, albeit a human universal, varies among people of different cultures. To test this idea, anthropologist Joseph Henrich journeyed to remote parts of the world to see how indigenous peoples, largely isolated from the industrial world, would play the Ultimatum Game. First on his list was the Machiguenga in Peru. He was shocked. These people were stingy—really stingy. They proposed a mean of 26 percent, with the most common offer only 15 percent. Even more puzzling, despite the meager offers, rejection was rare. Henrich was perplexed, thinking that after so much effort he might have blown the experiment. Later, he worked with Colin Camerer, an expert in game theory, to reanalyze the data. Still no error surfaced. It took a while for the two of them to figure out that the "stinginess" of the Machiguenga was caused by a factor familiar to anthropologists: unique cultural tradition. The Machiguenga people live in self-sufficient families that seldom appeal to outsiders for help. As a result, an offer from an outsider—no matter how small—is deemed a free gift, which they will gladly accept. Subsequent tests in fifteen small-scale, traditional societies yielded results that were all over the board. The mean offer ranged from 27 percent in the Tsimané of Bolivia to 58 percent in the Lamelara of Indonesia. The mean rate of rejection also varied significantly, from zero to as high as 40 percent in the Gnau of Papua New Guinea.[26]

Why do people in some societies—the Aché of Paraguay, the Lamelara—routinely offer 50 percent or more? These so-called hyper-fair offers are made for a reason, and, again, lifestyle reveals the secret. The Aché are headhunters, and the Lamelara, whale hunters. Men in both societies rely on one another during hunting trips. Boom or bust, they are all in the same boat. Once successful in hunting, however, they often have to deal with an excess of food. As food loses its normal value as a critical resource, people are less motivated to keep more than necessary for themselves. Hyper-fair offers, on the contrary, can serve for tightening bonds with team members, which is crucial in hunting. Probably just as important, they can also be used as a public campaign strategy for building reputation, raising one's social stature in the community—something that money cannot buy in preindustrial societies. Apparently, this logic is shared in the lavish potlatch feast of several Native

American tribes in the Pacific Northwest such as the Tlingit, Haida, Makah, and Nuxalk.[27] In these societies, competition for generosity ensues when food is plentiful and thus is not much of a concern for survival. Clearly, social and physical environments can modulate the intensity of selection. As a result, the sense of fairness varies among different ethnic traditions.

Cultural heritage aside, social, economic, and political institutions can also affect people's sense of fairness by altering the levels of expectation through mores, conventions, regulations, policies, and laws. People in democratic societies, regardless of their economic condition, tend to expect equality in resource distribution and social status.[28] As an even split is the common standard of fairness, the near 50 percent offer accordingly becomes the norm when the Ultimatum Game is played by people in industrialized societies. In the Machiguenga, however, people do not anticipate anything from anyone outside their families. Their sense of fairness appears unchanged from birth. This may be why their low percentages in both offer and rejection are close to those in young children in industrial nations. Here, the adage "the exception proves the rule" works well in unraveling the evolutionary myth of our fairness instinct.

Further evidence for the implication of DNA in our sense of fairness comes from a super nice Peter, who is so kind that others consider him sick. There is a rare genetic abnormality called Williams syndrome, occurring at a frequency of 1 in 7,500–20,000 births. Those affected by the condition lack a segment of DNA, which includes about 25–28 genes on chromosome 7. Some of these genes are involved in making brain proteins, presumably affecting brain functionality. Since chromosomes come in pairs, one from each parent, the effect of the missing genes may not be dire, despite some obvious physical and intellectual disabilities, so long as one of the chromosome pair is still intact. However, with only a single copy, instead of two, for each of these 25–28 genes, the volume of proteins manufactured under the instruction of these genes is significantly reduced. The situation is comparable to having only one instead of two production lines working in a factory. As a result, patients afflicted with

Williams syndrome have difficulty with tasks such as judging people, but they show an increased ability in areas such as verbal fluency and sociability. They are unusually generous and love social events—they are distinguished by having what are known as "cocktail party" personalities. Exceedingly unselfish, they will go to great lengths to help people. The reduction in these brain proteins may partially liberate their cautionary stance, causing them to be nicer, more friendly, and more generous than they should be, given the norms of our society. They are also much more altruistic, less likely to cheat, and more likely to reciprocate.[29] This abnormality provides a rare glimpse into the genetic foundation underlying our sense of fairness. It's ironic that the generosity in people with Williams syndrome is often distrusted by "normal" people simply because their levels of trust seem too high for our society.

The quest for the biological basis of fairness has led to a recent surge of revelations in neuroscience and molecular genetics. Advanced imaging technology has allowed scientists to pinpoint brain regions that respond to fair and unfair situations, often providing unexpected insights. For instance, as mentioned previously, two common types of justice are procedural justice (fairness in the process, or how fair the process is) and distributive justice (fairness in the outcome, or how fair the resources are divided). These two kinds of fairness markedly differ in terms of the brain response they provoke. Unfairness in the process is linked to higher activity in brain areas associated with social cognition—the ventrolateral prefrontal cortex and the superior temporal sulcus. Thus, people on the short end of the stick need to be *knowingly convinced* that inequality is actually fair because of their minor roles played in the process. (Here, the eloquence of organizational managers and company executives matters.) Unfairness in the outcome, however, elicits a higher level of activity in brain areas affiliated with emotions—the anterior cingulate cortex, anterior insula, and dorsolateral prefrontal cortex.[30] Hence, a sense of unfairness is often *automatically provoked* when one sees how much less he or she actually gets compared with peers. That's why the two camps—one emphasizing justice in the process and the other stressing justice in the outcome—often cannot convince each other through debate. Our brains perceive and process them as different issues, requiring different neural loops and behavioral responses.

Of equal interest is the finding that the decision to accept fair or hyper-fair offers, as we have encountered in the Aché of Paraguay and the Lamelara of Indonesia, can be traced to the insula and dorsal medial prefrontal cortex.[31] Furthermore, the ventral striatum and ventromedial prefrontal cortex are also implicated in people's aversion to inequality.[32] These results indicate that fairness is not one thing; it has several subcategories that evoke responses in different regions of our brains. Apparently, variations in our brain structure and response pattern may have a strong influence on our perceptions of fairness.

Putting variability in brain response aside for a moment, let's focus on the dorsolateral and ventrolateral prefrontal cortices. Together with the striatum, they are directly connected to and receive information from the dopamine neurons in the midbrain, as discussed in chapter 2. These structures along the dopamine pathway appear to be the axis of the neural network in our response to fairness situations.[33] This explains why fairness judgment and fairness action can be severed when the functionality of the right dorsolateral prefrontal cortex is disrupted by the magnetic field we discussed earlier in this chapter.

While some of the brain regions involved in our fairness response may sound alien, the hormones and neurotransmitters busily shuffling information in our brains appear to be the usual suspects. For instance, the arginine vasopressin receptor 1A (*AVPR1A*) gene is known to be involved in altruistic behavior. For example, three-and-a-half-year-old children with a variant (RS 327) of DNA in one of the promoter regions of the *AVPR1A* gene tend to be more "stingy" than those without it. Since promoters control how genes work, it appears that RS 327 either delays or reduces the expression of the *AVPR1A* gene.[34] For adults, a low level of serotonin, popularly dubbed the "happy hormone," can make us feel out of sorts and more likely to reject unfair offers in the Ultimatum Game.[35] Sex hormones—androgen and estrogen—are also involved in our fairness response in an interesting way. Among Chinese and Israelis, the estrogen receptor β gene (*ESR*β) affects the response in women but not men. Logically, we would expect that men might be correspondingly influenced by the androgen receptor gene (*AR*). But this inference is only partially correct: *AR* only affects Chinese men, not Israeli men![36] These sex and ethnic differences in the genetic underpinning of our fairness instinct mirror the hallmark of Darwinian evolution: variation, which provides raw materials for natural selection to work on. Similar sexual variations are found in another gene involved in the dopamine pathway (*DRD4*, short for the dopamine 4 receptor gene), which, together with several environmental factors, also affects our fairness response.[37] Here again, genetic variation is the wellspring of diversity in our fairness perception.

With waves of new discoveries—particularly in the fields of molecular genetics and neuroscience—the question of whether our fairness instinct is lodged in our DNA can be laid to rest. Viewing fairness as a human instinct is simply acknowledging its underlying genetic, hormonal, and neural mechanisms, entangled with our physical, social, and cultural environments. With this understanding, we can make much better sense of the ways our fairness instinct expresses itself and influences our perceptions and reactions in our daily lives.

Although reciprocity tends to be based on equal exchange, it's not always the case. Asymmetrical reciprocity, where one party gains more than the other, can still evolve. Consider several scenarios. Unlike in the Ultimatum Game, where no effort is involved, in social partnerships, one member of a team may contribute less than others to team efforts, and, if so, the split of the reward is generally expected to be uneven. For example, two partners invest $70,000 and $30,000 respectively in a joint business project. If they make a profit, a 7:3 split seems like a fair division. What about the salaries for two baseball players, one battling .310 and the other, .240, all else being equal? In these hypothetical examples, the parties contribute unequally to a joint effort, so to reward them equally doesn't seem fair. In such cases, the equality rule won't have much of a chance to emerge and evolve. To be fair, rewards must be proportional to the effort or investment each party contributes to the whole. This way, nobody will be rewarded for doing less. This general principle is known as the equity rule.[38] Ability and effort are two major elements in deciding the proportional partitioning of profits. Capuchin monkeys and chimps, as well as humans, can tie reward to effort, demonstrating their ability to apply the equity rule.[39]

The equity rule should have evolved according to the same logic as the equality rule discussed earlier. When the equity rule is used to justify an uneven distribution of resources, people's expectations are generally reset in proportion to their contributions. Therefore, fairness is still judged by the difference between actual payoff and expectation. By comparing the slogans "equal pay for equal work" and "to each according to his contribution," it's apparent that the equality rule is a special case of the equity rule, when the contributions between partners are equal.

Although the equity rule seems neat and tidy in theory, in practice, it can be difficult to apply, as there is often no objective and accurate way of tallying people's contributions. In applying the equity rule, most societies—from laissez-faire capitalism to welfare socialism—adopt some sort of merit system, based largely on ability and effort. However, merit is frequently a subjective matter that, under a wide range of circumstances, is not easily quantified. For instance, who deserves a higher salary, a productive researcher or a superb teacher in the same university? A meticulous quality controller or a creative designer in the same car manufacturing plant? Whenever and wherever merit becomes an issue, from governments to schools, from public to private sectors, and from military to civilian organizations, one can always question whether effort or talent is more important. This intrinsic ambiguity in merit assessment is a perennial source of debate and disagreement. It's a major reason why opinions differ wildly with regard to such hot-button issues as how much we should pay in taxes, how much of that tax revenue should be put in our welfare system, whether illegal immigrants should be granted the same benefits as citizens, and whether we should have a universal healthcare system.

Despite ongoing controversies over how to equitably distribute both benefits and costs, our evolved sense of fairness still lies underneath this sea of diverse opinions; it comes to the surface whenever the equity rule is egregiously violated. There have been numerous examples of this in recent times across the United States.

Before going on to examine some of these examples, we should be aware that most kinds of cooperation—joint ventures, teamwork, collaborative endeavors, collective enterprises—have both risks and rewards. The equity rule should apply equally to the downside as well. That is, the party who gets more when things go well should also take more responsibility or punishment when the venture goes awry. For the fans of the Los Angeles Lakers, when their team wins, Kobe Bryant, their star player, is often the man to praise. But when the Lakers lose, he is also frequently the man to blame. The equity rule, as leverage in resource distribution, can work both ways. In investment, both dividends and risks of a business are distributed among investors in proportion to the numbers of shares they hold. If this rule is violated, feelings of injustice and unfairness often result.

A recent case in point involved Carly Fiorina, the former CEO of technology giant Hewlett-Packard. Fiorina made a series of poor decisions, including the acquisition of the former computer maker Compaq. Needless to say, the company did poorly under her tenure between 1999 and 2005; its stock value was cut in half, from over $40 down to about $20. Nevertheless, when she was finally ousted, she received a $42 million settlement, including a severance package of $14 million. She found herself under fire from the business circle and drew a great deal of resentment from shareholders. The sentiment behind the hard feelings was apparent: giving her rewards, rather than punishments, for poor performance clearly violated the equity rule and was neither fair nor just.

If Hewlett-Packard could suck up the payment for Fiorina with its own money despite public disaffection, the AIG bonus controversy, as noted in chapter 1, was more emotionally charged. With the infusion of $182 billion in bailout money, the company was 80 percent owned by American taxpayers. In the public eye, AIG executives were, by the equity rule, liable for their excessive risk taking in business. At the very least, they were expected to take a prudent stance, owning up to some responsibility for the near collapse of the insurance giant. In some companies, executives savvy in public relations take a symbolic pay cut as a self-inflicted penalty to demonstrate humility. For instance, Citigroup CEO Vikram Pandit took a nominal annual salary of $1 and vowed to turn the firm around after the real-estate bubble burst in 2008. In 2009, however, this was not the case for AIG. On the contrary, while the company was still deep in the vortex of the financial crisis, AIG executives continued to spend lavishly, including $165 million in bonuses for their "performance" in the previous year.

Amid the public outcry, Treasury Secretary Tim Geithner appealed to the company to withdraw the bonus payouts. But the CEO of AIG, Edward

Liddy, sternly rejected the plea. Sandwiched between public uproar from the outside and instability, uncertainty, and low morale on the inside, Liddy, shouldering the mission of saving the company, had to appease both sides. He took a symbolic annual salary of $1 for himself while resisting public pressure against the bonuses. In a carefully crafted letter to Geithner on March 14, 2009, Liddy stressed two main reasons for why he would dole out the bonuses: "[The bonuses] are legal, binding obligations of A.I.G., and there are serious legal, as well as business, consequences for not paying . . . . We cannot attract and retain the best and the brightest talent to lead and staff the A.I.G. businesses [without them]."[40] If the first reason, though not unalterable due to the circumstance, made sense, the second was a total flop. Liddy clearly misread public sentiment. Angered by outright injustice, many Americans saw these self-styled "best and the brightest" as the culprits who toppled AIG. Instead of being rewarded, they should have been fired as punishment for their inexcusable errors and mistakes that had brought down the company, along with a large number of investors and, in the eyes of many, the US economy. "The American public's real objection to the bonuses—and the bailout—is not that they reward greed," points out Harvard political scientist Michael Sandel, "but that they reward failure."[41] The equity rule was turned on its head.[42]

A former CEO of Allstate, Liddy came out of retirement to take the helm of AIG in 2008 when the company was in deep trouble. Despite having nothing to do with creating the crisis, he was certainly aware of public sentiment. Some business and Wall Street executives were not. When executives from distressed automakers—GM, Chrysler, and Ford—asked Congress for bailout money, they came to Washington, DC, in private jets, as they were accustomed to doing on business trips. While using private jets only raised eyebrows when the economy was good, doing it when asking for corporate charity was a symbol of callous arrogance when so many people were losing jobs. Such a colossal stumble in public relations disgusted many Americans, who, though aware that the collapse of the auto industry would hurt autoworkers much more than the executives, would still rather see these companies go belly-up rather than give them another handout of public money. They wanted a way—any way—to penalize executives. This was why the government's decision to save the auto industry was unpopular among many

Americans, who were moved by this insult to their sense of fairness.[43] Again, the public fury was set off by the violation of the equity rule.

On January 27, 2010, President Obama, in his first State of the Union speech, was cheered by the audience when he made the rousing declaration, "If there's one thing that has unified Democrats and Republicans—and everybody in between—it's that we all hated the bank bailout. I hated it." After long, thunderous applause, he added, "I hated it. You hated it. It was about as popular as a root canal."[44] These punch lines were probably the most popular of the night. Indeed, if we could find one national issue that unites ultra-right Tea Party activists and ultra-left Occupy-Wall-Street protestors, it was the anger at bank bailouts. In reality, to the American public the bailout itself might have posed less of an issue than the deep resentment for rewarding failure. Behind the public ire was a broad sense of distrust, cynicism, unfairness, and injustice regarding corporate executives' abuse of American taxpayers' money. The ensuing political and economic malaise of recent years has shown how the fairness instinct has played out in public discourse, and, more to the point, it has unveiled the fairness instinct as a game changer in the American political arena.

This chapter has provided an abundance of evidence—from behavioral, pathological, neural, hormonal, and genetic studies—to support the thesis that fairness is indeed lodged in our DNA. This is why we often have a gut feeling for issues from minor traffic disputes to major political controversies when our sense of fairness and social norms of fairness are violated. Even more important in theory, though, is the concept of relative fitness. It is the key to understanding the evolution of fairness, the framework constructed in the previous chapter.

In concluding part 1, there is nothing I would put more emphasis on than relative fitness. A conceptual leitmotif that reverberates throughout the book, relative fitness is an entirely new scientific revelation in understanding a diverse range of major behavioral and psychological issues related to our fairness instinct. First on the list is a common yet still vaguely understood topic prevalent in our social lives: envy. It will take center stage in part 2.

Part 2

THE SURVIVAL OF THE MEDIOCRE

As a stabilizing selection force in Darwinian evolution, trimming off deviants from accepted social norms, fairness levels social and economic statuses among people. Envy, as an emotional expression of the fairness instinct, is exactly such a social selection force. By motivating people to take action to reduce the comparative advantage of envy targets, envy can even up rivals in social and economic status. Though broadly rejected in both Eastern and Western societies, envy is still a pervasive sentiment in a wide spectrum of human interactions from gift exchange and conspicuous consumption to the pleasurable feeling for the downfall of the famous, the wealthy, and the glamorous. Actions motivated by envy, as anti-intellectualism illustrates, can happen regardless of their dire consequences in our personal lives and our societies. The section uses a true story as a compelling narrative arc, under which a wealth of scientific information about envy, together with many comparative examples, is unfolded.

Chapter 4

LIFE IN A "RED-EYED" COMMUNE

When doing the research for this chapter, I was overwhelmed by numerous examples of everyday envy. Selecting the best one to illustrate my point became a daunting challenge. Offhandedly, my mother gave me a piece of advice, "Ask your aunt. She should have a good story to tell."

Born in 1950, my Aunt Jane grew up in East China in a farming community of about two hundred people comprising some thirty households.[1] The community, originally a clan, was converted to a commune after the Communist Revolution in 1949. It was also a place where I spent quite a bit of time in my childhood. Drawn by fond memories, I went back to the commune in the summer of 2011, with the mission of learning what Jane's story was all about.

A newly constructed six-lane freeway took me right to the doorway of the commune. Much had changed in the past two decades. Gone was the serenity, along with most of the familiar faces. Jane no longer lived in the commune. I met her at her son's apartment five miles away. Jane had only the one son, whose four-year-old daughter had a fever that day. Jane had volunteered to babysit her.

"Do you have a moment to spare?" I asked.

"Sure. The little girl is sleeping."

"My mom said you have a story to tell about your early years in the commune."

"Yes, but it's a long story," Jane replied, apparently eager to tell it. "I graduated from middle school at sixteen in 1966 and was sent to work in the commune for reeducation. It was Mao Zedong's way of teaching us 'good character' by laboring in the field."

Unlike its parallel in Stalin's Soviet Union, the Chinese collectivization was halfhearted. Although lands were seized by the government, some assets, such as houses and furniture, were still in private hands in the rural communities. The distinct signs of the "New Society"—an alias for Communist China—were that people worked together in the fields and shared major farming tools and products.

Midsummer was the busiest time in the region, as spring crops had to be harvested and fall crops planted on the same field in the narrow window of one month.[2] During this time, everyone in the commune toiled in the rice paddies from dawn till dusk, save for breaks for food and rest every three hours.

"Didn't the commune open a community kitchen in the summer?" My memory was blurry.

"Yes, it was Communist and saved time. All households could dine without preparing their own meals," Jane smiled. "The community kitchen and my room were only separated by thin planks of wood. The aroma in the kitchen seeped through the wall and woke me up every morning."

Jane's commune was self-sufficient, as were most others in China at the time. Peasants sold a quota of rice to the government at a fixed price and all extra grains and produce were distributed to the households in the community. Watermelons, for example, were distributed by a random procedure. They were first sorted into piles according to size and quality, then labeled and given to each household based on numbers drawn from a straw hat. I told her I still remembered how watermelons were distributed fairly.

"You still remember what you did! You loved melons, often rolling them before they were allowed to be taken home. And you didn't have good table manners. After every melon feast, I had to wash out the stains on your shirt." After a few chuckles, she became more thoughtful as she contemplated how hard life had been. "At that time, we were all very poor. The families in the commune were pretty much all alike. We had pickled vegetables for meals almost every day. Salted fish were considered a delicacy. Pork and chicken were mostly saved for special occasions. Candies and fruits were rare and often eaten quietly and secretly, in case others saw. New clothes were only for the spring festival, the lunar New Year in China. They were worn for years and mended repeatedly until they were full of patches. They were faded, dull in color—blue, gray, black—and similar in style for all in the commune, young and old, men and women. No one really stood out in any aspect of everyday life."

"Recently, I read a book about how ordinary people live without envy in North Korea," I told Jane. "North Korea is a Communist nation that still uses a system of planned economy with little private property. It's quite like the China of your youth."

"No envy in North Korea?" She was surprised. "But that was not true of my commune."

"I don't believe it, either," I agreed. "My mom said you somehow became a target of envy in the commune. What happened?"

"In the beginning, I was full of youthful energy, ready to devote myself to building a prosperous New Society. I was passionate about work and fast at it. Whenever commune members worked together, regardless of harvesting, planting, or weeding, I did more than others. I had no idea my effort made others look bad, as if everyone else was slow and lazy on purpose. The leaders praised me for my diligence, but I didn't know they weren't sincere. Before long, several people began to grumble and comment bitterly behind my back. Although the leaders gave me hints to slow down, I didn't get them at first—until the work points I earned were often equal to those who did less. I was hurt."

Work points, Jane explained, were used to assess how much one contributed to the joint effort, which would be exchanged for some of the cash from grain sales. They were assigned daily to each person according to a merit system. After conversion, a full day's work in Jane's commune earned a strong laborer just over half a yuan,[3] the equivalent of seven cents in the US. Men were normally considered "full laborers," usually earning about one work point per day. Deemed weaker and burdened by female cycles and child bearing, women typically earned around half a work point each day, even when they performed the same amounts and types of work as men. Points from team work were awarded through a democratic process: leaders proposed how many points each person would receive, and all members voted on it.

"I was too young to understand how complex the system was at the time," I added. "How many work points did you earn in a year?"

"A lot! Even though others were not fair to me, I still earned 260 points in my first year, many more than any other woman in the commune. In fact, I set a new local record, and my accomplishment was broadcast in several neighboring communes."

"Apart from your high work efficiency, my mom told me you were also artistic and had a gifted voice."

"Yes, I was. From the moment I came back to the commune, I busied myself in theatrical plays, basically modern Beijing operas, such as *The Red Lantern* and *The White-Haired Girl*, endorsed by the government for the purpose of propaganda. I performed not only because I liked the performing arts, but also because I wanted to help raise the working morale of people for socialist construction. Even if I was totally exhausted after a day's manual work, I still dragged myself to practices and performances in the evenings."

Indeed, for the locals, who didn't have electricity until the early 1970s, Jane brought joy to the otherwise dull and dreary farming community that had virtually no other entertainment to brighten their daily routine.

"I still remember that, on the nights of the shows, folks would come to see your performances in droves and crowd around the makeshift stage, on the flat ground normally used to dry grains." For a child like me, it was always a spectacle to experience such large crowds.

"Some fans followed our performances in different communes and watched the same shows again and again, dozens of times," Jane added. "However, a few in my commune became envious. Perhaps I was too popular in the region."

"In other words, you were too good, too ambitious, too outstanding for some in your commune. But you were too naïve to pick up the hostility toward you, right?"

"Actually, I was aware of it, but I was slow to figure out what I did wrong. I put all my efforts toward supporting the Communist cause. How could my selfless dedication have been hated? For a while, I was quite troubled by how unfairly my work was assessed, until a friend of mine told me how others felt."

"So you learned a hard lesson."

"An important life lesson indeed. As an old Chinese proverb says, 'the bird whose head pokes up gets shot.'"

"Did you know there is an English expression with exactly the same meaning?" I added. "'The nail that sticks out gets hammered.'"

"I didn't know that. Had I known what they were thinking, I still might not have bowed to the pressure of some noisy commune members. I saw myself as strong as Tiemei in *The Red Lantern*."

The Red Lantern is a drama cast during the Sino-Japanese war of 1938–1946. It's about three members of a Li family: Yuhe, the father in his forties, Grandma in her sixties, and the seventeen-year-old Tiemei, a name that literally means "iron plum blossom." Yuhe is a spy for Communist guerrillas and is given the task of delivering a secret telegraph codebook to the resistance, with the Japanese army in pursuit. After capturing Yuhe, but failing to find the codebook, the Japanese execute Yuhe and Grandma, but leave the shaken Tiemei alive so they can extract information. Tiemei, instead of being brought to her knees, assumes her father's role. Courageous and quick-witted, she succeeds in delivering the codebook to the Communist guerrillas.

"Was Tiemei your role model?"

"Yes, for many years she was an inspiration to me."

"What did you think when those in the commune turned against you?"

"I simply wanted to leave. I continued to work on theatrical plays whenever there was a break. The opportunity eventually came in 1970, when my program won an award in a county performing arts show. The people from the county performance troupe wanted me to join them."

"So you thought your dream of becoming a professional performing artist was finally coming true?"

"Yes! I was so excited that I couldn't sleep that night." Jane's eyes lit up, then dimmed. "But the commune party boss intervened. He appealed to a higher authority to withhold my registered residence. If my residence could not be transferred, nobody would take me. I would be stuck in the commune. Moreover, the party boss lobbied the performance troupe to dismiss me. In the end, I was forced to come back to the commune after just three months in the troupe."

"How did you feel at the time?"

"Disappointed, dejected, angry, I guess. It was very hard for me to accept."

"How did you recover from such a huge letdown?"

"Tiemei inspired me. She chooses to fight, so why shouldn't I? I decided to fight against the injustice. I told myself, 'I must not stay in the commune, as they wish me to.' Also, with so many envying me, I didn't think I had a future in this tiny commune. I began to seek opportunities elsewhere, a place with fewer envious red eyes and more artistic activities."[4]

"So you left the commune for a teaching job, if I remember correctly. My mom was a teacher. Did you talk it over with her?"

"Yeah, I did. But when I saw your mom working so hard at school every day, in the beginning I felt a little intimidated by teaching."

"What made you change your mind?"

"Two things. One was that the manual work was just too tough for a woman—any woman. Sometimes I wasn't allowed to stop laboring even though my nose was bleeding. I hated it. The other was the envy. It was unpleasant to know people were looking at me like that." Jane's tone turned hopeful. "When I got to the school, I found that I liked teaching. Working with the children made me feel good. The joyful songs, dances, and laughter in the classroom reminded me of my school years not long before. Also, there were no shortages of artistic activities and music instruments I could play."

Despite Jane's desire to teach, how could she—only a middle-school graduate—have been qualified as a teacher? In the tumult of the Cultural Revolution, manual work was glorified as a builder of proletarian character. Illiterate or semiliterate farmers, factory workers, and retired military personnel assumed leaderships in schools, colleges, hospitals, presses, and other cultural institutions. Intellectuals, meanwhile, were silenced and sidelined. Although teaching was still considered a desirable job, qualified teachers were few and far between. As a consequence of anti-intellectualism, many schools were short of teachers to cover their classes. Since, at the time, most Chinese had little education, Mao considered those with four or five years of elementary schooling to be intellectuals. Thus, Jane was qualified as a teacher.

"I remember you once taught at my mom's elementary school."

"Yes, it was the first place I taught, with over 1,200 students, just three miles away from my commune. As a substitute teacher, I covered a wild range of subjects, from music to physical education."

"Your long braid was quite a spectacle at the school," I laughed. "Why did you keep your hair that long?"

"I needed it for performances, because my favorite role, Tiemei, has a long braid and we didn't use wigs." Jane paused for a moment, then continued more brightly. "I had to carefully plait my hair every day. It dangled behind my back, all the way to my waist. The schoolchildren had never seen anyone with such long hair; they were fascinated. A few naughty boys snuck behind me at times, jerked my braids, and ran away laughing."

"It was fun." After the interjection, I thought I should change the topic. "How much did you earn?"

"My salary was eighteen yuan (about US $2.50) per month. Not much, but more than I earned working on the commune farm."

"But why did you quit teaching?"

"I didn't quit; they called me back to the commune," she corrected. "Just as life was beginning to look up for me, I forgot that the very act of leaving the commune was enough to vex people. Some in the commune felt it was unfair for me to avoid physical work in the field. They argued that even though a person had some education, she should not be treated differently from everyone else. They believed that regardless of the need for teachers and my personal preference, farm work was at least equally important. In their minds, without manual work, my soul might be corrupted by evil bourgeois ideologies. After much discussion, members of the commune concluded that because everyone else labored in the rice paddies, I should be summoned back. The party boss came in person and urged school authorities to fire me, so I was thus released. With nowhere else to go at the time, I returned to my commune. I felt terrible. I lost my enthusiasm for work."

"Returning must have been difficult."

"It was. Moreover, it was obvious I was being singled out. I was often assigned the hardest and filthiest work—building reservoirs, dredging mud from river bottoms to enrich rice paddies, and, the most unpleasant of all, scooping pig manure for fertilizer with my bare hands. Some of the work was shunned even by the strongest of the men. In the winter, my hands and feet were full of cracks, and I got frostbite all over them; often they were bleeding, but I still had to work."

"How did you endure the punishing manual work for that long? Did you complain?"

"Again, Tiemei inspired me. In comparison, what did I have to complain about?" Jane sighed. "I endured the hardship for another reason. I might have an opportunity to go to college."

How was it possible for her, a middle-school graduate, to have an opportunity to go to college? In the mid-1970s, the bleak consequences of the Cultural Revolution became evident, resulting in widespread economic hardship. Mao was senile and seldom appeared in public. Deng Xiaoping presided over the daily running of the central government. Schools were back in session. Colleges and universities reopened and began to enroll new students. However, they took only "good students" born to proletarian or poor peasant families who had demonstrated their dedication to the Communist cause. Such "red" college students were selected based not on their academic achievements but on recommendations from their local leaders. Jane was lucky to be among the few who met these criteria.

"Even the commune party boss promised to give me the opportunity if I continued to do well," Jane went on. "I was hopeful about the prospect—I would be able to study the performing arts in college. Life can be so dramatic!"

"But your college dream never materialized, right?"

"Dream? More like a nightmare! It turned out that the promise of sending me to college was a gimmick concocted by the party boss. He intended to get me hooked for the time being. The single-person quota eventually went to the son of a higher local party boss. All my lifetime dreams that seemed so close to being realized were now shattered."

"What happened?"

"I'd been suspicious about why I had been so 'unfortunate.' After the college drama, I figured out why the party boss was so eager to keep me in the commune. In addition to the envy I'd aroused in some members, there was a deeper layer of complication. As you may know, I was considered very pretty at the time."

"My mom told me you were well-known for your beauty in the region."

"Your mother was, too," she added. "Over the years, I had many suitors, but I turned them all down. Feminine attraction was a major factor in the performing arts, though this was not acknowledged in public. A female artist, once married, would find it a lot harder to continue her career. So I decided to put off my marriage indefinitely."

"I bet the party boss noticed it."

"Not only that. He actually fell for me. Though tall and masculine, he was blunt and boorish. With his power in the commune, however, he thought he had a chance. He wrote me a courtship letter, but it was full of basic errors, not to mention bad wording and grammar. This turned me off even more. I shot his marriage proposal down at once. Though discouraged, he hoped I might one day change my mind. But if I left for another place to take a job in a school, join a performance troupe, or attend a college, his hopes of marrying me would be dead. This was another reason why he tried everything possible to keep me in the commune—including cajoling me into believing he would send me to college and forcing me back from the performance troupe. Backed by a few envious, noisy members in the commune, he had little to be afraid of. When I disobeyed him, he retaliated by assigning me the toughest and the dirtiest work. For the same reason, I was unable to join the Communist Youth League."

The Communist Youth League is organized by the Communist Party. Membership in the league brings opportunities for social and political advancement. Although Jane was unaware of the political privileges associated with the league at the time, what she was denied was well-deserved recognition of her excellence at work. Envy and jealousy spelled doom for Jane's dreams.[5] What are the chances envy could motivate the party boss, the Orwellian Big Brother reigning in this little kingdom of the commune, to pursue and persecute her so persistently and brazenly? Jane had not realized the core of the problem until this petty party boss had singlehandedly wrecked all her lifetime opportunities. It was a devastating moment.

"So, regardless of your obvious no-win situation against mediocrity, envy, and jealousy, you seemed determined to fight to the end."

"Yes. I left my village again to take a teaching job in a remote town. I was pretty sure I would never come back to the commune."

"Did the party boss come after you again?"

"He did, and with vengeance instead of jealousy. But this time, the school authorities were on my side. They rebutted him and sent him back."

Our conversation went on for another hour. With an amazingly detailed memory, Jane told me much, much more about her stories, stopping only to wipe away tears.

"Do you still sing?" I asked.

"Only occasionally." Jane said she goes to karaoke bars with friends and relatives once in a while. These are popular venues for artistic expression and entertainment in Asian countries. The dreams of her youthful years were still alive.

While Jane was talking, her granddaughter woke up. She was just as beautiful and artistic as Jane had been. Jane held her and began to hum a lullaby. Regardless of their vast age difference, the two somehow looked alike. Jane's genes, talents, ambitions, courage, and hopes appear to live on in the little girl.

After my interview with Jane, my mind became preoccupied with several questions. Why did she become a target of envy? In a community where people lived in abject poverty and had so little to compete for, and so little to gain from envy, why did some still contract the "red-eye" disease? It will take an entire chapter to answer these questions and explore their serious ramifications in our everyday lives.

Chapter 5

THE SURVIVAL OF THE MEDIOCRE

In the previous chapter, while Jane understood the literal meaning of the Chinese proverb "the bird whose head pokes up gets shot," she was unable to grasp its implications in her own life. What can we make of the hostility toward Jane from some of her peers? Revisiting the games discussed in chapter 3 might provide some insights.

In both the Ultimatum and Dictator Games, we have learned that people, when given an opportunity to play Peter, the proposer, tend to be more generous than rational choice theory predicts. The discrepancy results from a human mind that may be—at least in part—tuned to long-term gains from repeated cooperation rather than focused solely on maximum profit in the immediate context of a one-shot deal. There is, however, another force that may push Peter to be more generous in games as well as real-life situations.

In most Ultimatum and Dictator Games tested around the world, volunteers play the games with only a single partner who is unknown to them. Real-world interactions among people are much more complex, often with three or more parties involved. To add a little more fun and realism, put a third person, Tim, into the Ultimatum Game. What will Peter, the proposer, do when his transaction with Richard, the recipient, is visible to Tim, the third-party bystander, who has some power to intervene? What will Tim do when he sees Peter stray from the path of fairness? Two Swiss economists, Ernst Fehr and Simon Gächter, made a notable contribution toward answering such questions. In a landmark experiment, the two researchers drilled deep into our sense of fairness when we are put in Tim's shoes.[1]

In the experiment, Fehr and Gächter recruited 240 university students in Zurich. To start with, each student was given $20.[2] The students were then broken into teams of four to play an investment game,[3] consisting of six consecutive rounds. In each round, a player was allowed to invest a certain amount, between $0 and $20, in a team fund. For each dollar invested by any person in the team fund, the return, donated by the researchers, is $0.40 for each person in the group (or $1.60 for all four members combined).[4] At the end of the game, the team fund was evenly split among the four members. In case cliques developed, Fehr and Gächter concealed the identities of the players and shuffled them so that no two were in the same group twice. This guaranteed no trust, reputation, or reciprocity could possibly be involved among the players.

If you are a player, will you, being lured to such a large potential return, invest all you have, that is $20? You may expect to bag $32 personally,[5] if—and there is a big if—all three of your teammates throw in all of their initial funds as well. However, your teammates may also all choose to be free riders, contributing nothing to, yet taking from, your investment. If this worst-case scenario happens, you will end up with only $8 from your $20 investment—a 60 percent loss—after the first of the six rounds in the sequence. Meanwhile, your three selfish teammates will each gain a profit of $8—a 40 percent return—from *your* investment. Facing this dilemma, what did the participants do? Fehr and Gächter found that players invested an average of about $10 in the first round. In spite of the high return, the amount invested dwindled to a mere $4–$5 by the sixth round, as the initial optimism of participants faded to cautious cynicism over time. Given this rate of decline, if allowed to continue, the game may not go much further before each player's contribution drops to zero. The moral: free riders are a menace to cooperation and a drain on esprit de corps.

Next, Fehr and Gächter let the volunteers play six more consecutive rounds of the same investment game, with one difference—players each could use up to $10 to punish the uncooperative behavior of their teammates. For every dollar the punishers put in, the punished had to pay $3 as a fine. If no player used any money for policing, the game would be the same as the previous six-round series. However, if a player sacrificed the maximum of $10 allowed for punishing, the punished would wind up paying $30. With this slight modification, the results were markedly different, as if more than just this one simple condition had changed. In five runs of the six-round series (that is, a total of thirty interactions), 84.3 percent of the players punished their teammates at least once, 34.3 percent over five times, and 9.3 percent more than ten times. As we would expect, the punishments were mostly (74.2 percent) done by those who contributed more than the average against those who contributed less than the average. With policing allowed, teammates became eager to cooperate. The amount of money invested by each member jumped by over 30 percent, from about $12 in the beginning to about $16 (only $4 short of the maximum) by the end of the six-round series.

These results tell us that people are willing to punish unfair behavior at their own expense. This is why most of us are willing to pay taxes to support law enforcement. In other words, we are ready to sacrifice some of our own benefit to ensure that fairness is maintained and cooperation is encouraged. Fehr and Gächter gave such punishment a name that might seem paradoxical: altruistic punishment. It serves as a crucial means to uphold Locke's social contract, by which people can exchange some of their natural rights for enjoying the advantage of living in a civil society.

The role of altruistic punisher is exactly what Robin Hood assumed in medieval England by robbing the rich to give to the poor, keeping unfairness and injustice at bay. The prevalence of such willing punishers in human societies makes people mindful of the consequences of veering too far from the accepted norms of fairness. This explains why, in the Ultimatum Game, the presence of the third party, Tim, tends to make the proposer, Peter, much more generous than when Tim is absent. The immediate payoff for Peter's tactical generosity is a lowered risk of penalty for his greed. This explains why Peter is nicer than he may want to be without such a string attached. But how can controlled scenarios like these give us insights into the kind of hostility experienced by Jane?

The role Tim exerts in policing behaviors that deviate from social norms is as intriguing as it is important. Such a force, in the jargon of evolutionary biology, is called *stabilizing selection*. As the name indicates, stabilizing selection maintains the status quo by boosting those in the middle while eliminating outliers. It works like pruning—clipping off side branches helps the trunk grow faster. This type of selection can prevent behavioral, physiological, or morphological traits—unfairness in our case—from drifting too far from the average.

One textbook case of stabilizing selection is human birth weight. The optimal size for a newborn is 6–8 pounds. Underweight infants may suffer from poor development, whereas overweight infants may experience complications in the birthing process. Only when modern biomedical technologies became widely accessible did newborns outside the optimal weight range (and their mothers) have fewer problems. The broad availability of Cesarean section is a relatively recent development, so it has had little chance to have an evolutionary effect on human birth weight.[6]

Likewise, stabilizing selection retains human facial features. Straying too far away from symmetry, for instance, will make a person unattractive as a potential mate, ultimately leading to a loss in fitness. Quasimodo, one of the most successful characters created by Victor Hugo, is, despite a kind heart, an unpopular choice in the mating market because outward physical traits carry a great weight in human mate choice. Once again, stabilizing selection prevents those with distorted facial features from becoming common, a reason why Frankenstein's monster remains a fictional character.

Stabilizing selection is a quiet, tough laborer, tirelessly toiling on dull routines. Usually, its contribution is unnoticed until it takes a day off. The Mexican tetra, or blind cave fish, for example, has lost its vision, as sight has little survival value in the darkness of deep caves. Here, the selection force that keeps the eyes functioning is off duty, and mutations that blind the fish can be passed on. Moreover, the loss of vision frees the fish from the physical demands and energy required for eyesight. Similar to reducing utility bills to save money for other uses, the mutations that disrupt vision enable the fish to channel the energy normally used for eyesight to other vital functions.[7] Thus, over time, such mutations can build up until all the fish in a population are taking advantage of being blind, fulfilling the evolutionary imperative for animal organs: use it or lose it.

Stabilizing selection, which keeps most of our body features within a narrow range of variation, also maintains a large majority of species' stereotypic appearances as if they have never changed. Even Carl Linné, perhaps the greatest botanist of the eighteenth century, was fooled by the apparent stability of species he observed. Not aware of the distinct morphologies between the sexes separately shaped by stabilizing selection, he made the innocent mistake of designating male and female mallards with two species names, *Anas boschas* and *Anas platyrhynchos*. He would have easily avoided this blunder had he paid a little attention to the activities of male and female mallards in the spring.

In Fehr and Gächter's study, altruistic punishment is precisely such a stabilizing force. It constrains people's behavior, promoting cooperation by punishing those who deviate too far from the accepted norms of fairness. This can be seen broadly in hunter-gatherer societies. "People like Bushmen or Pygmies gossip incessantly and are highly judgmental," writes ethnographer Christopher Boem, "and group opinion is something to be feared because moral outrage can lead to ostracism, expulsion from the group, or even execution."[8]

In industrial societies, likewise, our penchant to police can be seen in nearly all aspects of our lives. As a result, extraneously imposed sets of rules, regulations, and policies have been meticulously crafted and implemented to control abuses. In American society, as in many others, judges, referees, and umpires are screened by and subjected to rules that ideally insulate them from potential biases in courts, sports, and contests, where fairness counts. Even Little League baseball or children's soccer events in small towns require careful reviews of the birth certificates of the participants. Nevertheless, arguments, suspicions, and accusations of favoritism and cheating may still arise, especially in controversial calls.

The role of stabilizing selection becomes more interesting when our third-party bystander, Tim, is introduced into the Ultimatum Game. Given that he is human, he is hardly free from self-interest. Assume he is a colleague of the other two players, Peter and Richard. What will Tim do when he sees Peter and Richard benefit from the game? Granted the power to intervene, he may be happy if neither Peter, the proposer, nor Richard, the recipient, gain. By blocking the profitable transaction between Peter and Richard, Tim can stay even with them in terms of relative payoff.[9] This explains why independent auditors tend to be more willing than they should (if they only focus on their own profit) to reject agreements between investment managers and investors, even at their own expense.[10] This is a case where natural selection for gaining competitive advantage reveals its dark side: *while favoring a person who does better than his peers, it also rewards him if he can prevent a peer from succeeding*. Thus is born an adaptive emotion known as envy, which motivates actions intended to cut down the advantages of social peers. *Envy, therefore, is a key component of the stabilizing selection that prevents others from doing better.* It is the ultimate reason that Jane was persecuted for standing out. Behind the envy was the fear that she might do so well as to break out of her fishbowl community. Apparently, the borderline between envy and concern for unfairness is fuzzy.

This analysis makes us appreciate Fehr and Gächter's study even more: though small in scale, it unveils the hidden instinct behind our willingness to police for fairness. Yet too often, such social forces can "misfire," turning against those who excel, as happened to Jane in her farming community. How does this drama play out in our lives? And, more important, what can we make of the motivation behind this common phenomenon?

Assume Tim owns a posh, thriving restaurant frequented by the glamorous and wealthy. He is concerned with his own image. But he is short—unusually short—to a degree that his physique, at first sight, is unimpressive. Given the authority of ownership, what would be his criteria for hiring waiters and waitresses, if unrestrained by antidiscrimination laws? Dwarfs, perhaps. For the sake of his own image and self-esteem, he may opt to pass on those who are tall and handsome. Such a Tim is not a real person, of course. (He is, however, a fictional character living in the Chinese classical novel *Heroes of the Marshes*, written around the fourteenth century.) Yet being unreal makes this Tim compelling, for there is no shortage of real-world examples in our experience.

A friend of mine, a teacher in a liberal arts university, once complained that his department chair—call him Prof. B—had little vision regarding the issue most interesting to him: faculty salaries in his department. Prof. B took every measure within his power to ensure that nobody else's salary exceeded his own. After being appointed the interim dean of a college, Prof. B, with his cynical belief that educational and research initiatives were often excuses for faculty and staff to make extra money, minimized spending by slashing financial resources for existing programs and rejecting new proposals whenever possible. His successor was dumbfounded when she discovered that he had been able to "save" so much from the budget in his short tenure of two years.

Of course, Prof. B's salary couldn't be compared with those in business or high-tech companies, especially before he was promoted to interim dean. Moreover, the average salary of the department under his rein was slightly below the university average. If Prof. B were economically rational, what he should have done is exactly the opposite: lead the department in lobbying the university authorities for a departmental salary hike. Had he done so and succeeded, everybody in the department would have benefited. Near the age of retirement, he himself, though, might not have been the person who gained the most.

Why did he take a course of action that, from an objective perspective, was detrimental for himself? The answer is evident in light of relative fitness: what mattered for him was not his absolute payoff, but his relative advantage over his most relevant peers. If everybody in his department got a raise, he would be *relatively* worse off. Subject to the evolutionary curse of relative fitness, his vision was blurred by comparing himself to the small circle around him rather than to the society at large.

Such cognitive bias is known as a "hedonic treadmill," where one is engaged in the delusion of self-advancement, a misguided indulgence inspired by the promise of relative advantage in economic or social status among peers. If falling behind, one may experience stress even though, in the absolute sense, one's life is comfortable.[11] This is a paradox branded *subjective well-being* by economists and psychologists.[12] "Subjective" refers to the feelings provoked by comparison.

To better understand the role of subjective well-being, psychologists Sarah Hill and David Buss asked 205 people what kind of world they preferred: one where they are attractive and have a good income but with people better off than they are, or one where they look plain and have a lower income but they are better off than everyone else. Most participants (84 percent of men and 88 percent of women) preferred the latter choice, where, despite a lower absolute income, they would be in a superior position from a relative standpoint. Additionally, women, more than men, were more likely to choose a world in which they are plain looking but more attractive than their same-sex peers.[13] This demonstrates how the hedonic treadmill works. Our preference for relative advantage illustrates a point crucial to understanding evolutionary processes that we first encountered in chapter 2: *natural selection is a localized force, operating in reference to peers, especially those in the immediate environment.* Likewise, the evolution of fairness is first and foremost a localized process, best mediated through face-to-face interactions dictated by relative payoffs.

A downward comparison, as illustrated by Prof. B in the previous example, can boost one's self-esteem. Outperforming others may lead to feelings of pleasure, satisfaction, and dominance. But what if the tables are turned? For instance, Harry drives a Honda Accord LX whereas a coworker owns a Lexus SUV GX. Harry has a moderate house in a middle-class residential neighborhood, whereas his childhood friend owns a large waterfront mansion in a gated community. Harry has a hard time finding a girlfriend, whereas his step-brother is always surrounded by gorgeous young women. Harry is less capable and efficient than his talented coworker Jane. What will he feel under these circumstances? Most likely, envy.

Though an unpleasant emotion, envy is nonetheless common when a person lags behind in a comparison. Philosophers were mindful of this centuries, if not millennia, ago. Francis Bacon stated it plainly: "Envy is ever joined with the comparing of a man's self; and where there is no comparison, no envy."[14] For Kant, envy is "a propensity to view the well-being of others with distress." It "aims, at least in terms of one's wishes, at destroying others' good fortune."[15] The modern definition of envy, as psychologists Richard Smith and Sung Hee Kim put it, is "a pain felt toward (or caused by) another's good, good fortune, prosperity, achievements, or favorable circumstances."[16] Envy, therefore, is a negative emotion, often evoked by *perceived* unfairness, when a person is outcompeted in wealth, status, talent, popularity, intelligence, or other capabilities or attributes.[17]

Envy is about rivalry, real or perceived. An athlete may envy a better performer in the same sport; a scientist may envy a more brilliant peer in her field; an unsuccessful plumber may envy a competitor whose business is thriving; a plain-looking woman may envy a beautiful one, which led Arthur Schopenhauer to claim "pretty girls have no friends of their own sex."[18] This may explain why there are so many "blond jokes" in America. Schopenhauer, by the way, had no qualm about practicing envy. In 1820, he wanted to deliver his lectures at the same time as his more famous rival Friedrich Hegel at the University of Berlin. When authorities decided otherwise, Schopenhauer resigned his post at the university.[19]

Outside the arenas of direct competition—at professional levels, in the workplace, among friends and relatives, or in the immediate neighborhood—envy can attenuate or even disappear. This was already clear at the dawn of the Enlightenment. It was obvious for Hume that "a poet is not apt to envy a philosopher, or a poet of a different kind, of a different nation, or a different age."[20] For Spinoza, "no one envies the virtue of a person who is not his equal."[21]

Envy is a complex emotion, so subtle that it may be mingled with a score of other emotions including a sense of inferiority, sadness, shame, humiliation, self-doubt, failure, helplessness, ill will, hatred, and resentment.[22] People often experience envy when they do not get what they want, a situation that may also elicit disappointment, frustration, and anger. Given an envy-provoking scenario, even six-to-eleven-year-old children can articulate envy-related emotions such as sadness and anger. Though children are susceptible to envy, they have a low level of malicious ill will,[23] perhaps a result of their inability to take the same sorts of actions as adults.

Modern imaging technology has enabled neuroscientists to piece together a rough map of the emotions tied to envy in the brain. Take a look at table 2, which lists some of these findings.[24] Do you find anything interesting?

Table 2. Some brain regions and their corresponding functions related to psychological and behavioral responses connected with envy

Brain region	Function/response
Nucleus accumbens	Schadenfreude (see below for explanation)
Medial prefrontal cortex (MPFC)	Social comparison; guilt; embarrassment; moral transgressions
Amygdala	Social comparison; aggression
Dorsal striatum	Aggression; Schadenfreude
Insula	Unfairness; empathy
Temporal pole	Embarrassment; moral transgressions

Modified from image in J. E. Joseph, C. A. J. Powell, N. F. Johnson, and G. Kedia, "Functional Neuroanatomy of Envy," in *Envy: Theory and Research*, ed. R. H. Smith (Oxford: Oxford University Press, 2008), pp. 245–63.

The answer: the brain centers associated with envy have a profound connection with the limbic system, including the amygdala, the headquarters of our emotional responses related to anger, fear, pleasure, sexual arousal, and long-term memory. Now we know why envy is often such a gut-churning emotion: it is linked to our basic instincts.

What purpose, then, does envy serve? According to Hill and Buss, it's an early warning system endowed to us by natural selection. By evoking an intense suite of interlaced emotions, it alerts us that we are being outcompeted by our peers in obtaining critical resources for enhancing our own fitness. It motivates the envier to take actions to avoid falling behind, either by catching up with or bringing down peers. Hill and Buss tested this hypothesis by asking people to recall past incidents when they felt envy. It turned out that both men and women experienced envy when they were in direct competition with and outdone by social peers in any of the categories measured: personal attractiveness, gifts from lovers, sexual experience, and attractive mates. Women, more than men, felt envy when their same-sex peers were more attractive than they. Compared with women, men became more envious when they were beaten in sexual experience by other men.[25] These results are telling in light of sexual selection: women's fitness tends to be staked on the quality (including critical resources) of mates, whereas men's relies on the quantity of mates—a biological principle known as Bateman's rule.[26] Put simply, compared with women, men care more about the quantity than the quality of mates.

While this explanation of its adaptive value makes evolutionary sense, envy can still be an unpleasant experience, as it may linger for a prolonged period of time, causing obsession and emotional pain. Its debilitating effects are often overlooked, as in the case of the real-estate business. Contrary to the motto "Location, Location, Location," the worst house in a wealthy neighborhood is, in fact, the most stressful place to live for a status-sensitive person. Our impressive neighbors' houses can be silent yet persistent reminders that we are the "little guys" on the block. By the same token, wealthy investment bankers may see bonuses and compensations as badges of their success and status in a company. Even though they may be rich in the absolute sense, driven by a perceived sense of fairness, they may still feel sulky in regard to colleagues who earn more.[27] These simple examples, backed up by a large body of scientific evidence, support our intuition that happiness is not assured by absolute affluence.[28]

At a deeper level, envy is a source of mental anguish, a notion common to observers of the human condition since the time of Socrates and Aristotle.[29] Its effects hurt enviers by lowering their self-esteem and perceived worthiness with a torrent of negative feelings: pettiness, mean-spiritedness, inferiority, misery, resentment, rage, and a habitual attitude of "sour grapes."[30] Social disapproval torments them further, for both Eastern and Western societies treat envy as a malicious emotion. In the Christian tradition, for instance, envy is a sin punishable by immergence in the freezing waters in hell. Feeling shame that can't be expressed in public, or even among close friends, an envious person is often left alone to cope with a deluge of negative emotions. This makes envy harder to overcome, since the envier feels unable to turn to outside help. Consequently, people in societies with a steep dominance structure tend to suffer more health problems, such as cardiovascular diseases and reduced immunity from persistent mental stress, as well as shorter life spans, than those in more egalitarian societies.[31]

Despite the negative physiological and psychological effects, not all envy is bad. Benign envy can prompt people to put more effort into their work in order to earn what they feel they too deserve or are entitled to, such as money, cars, houses, honors, promotions, and higher status. For this reason, some psychology-savvy American companies tacitly tap into the envy factor to boost the performance of their employees.[32] Likewise, people with below-average incomes are more motivated to use the lottery as an opportunity to improve their low economic status. They buy twice the number of scratch-off lottery tickets as those whose incomes are around average.[33]

Manufacturers, too, take advantage of benign envy to market big-ticket items. The strategy is highly successful today in Asia, which dominates the consumption of luxury goods in the world. In Tokyo, for example, over half of young women own some high-end products from Burberry, Luis Vuitton, Gucci, Prada, or Chanel.[34] And China is not behind. Apart from the high-end brands, in recent years, Coach handbags have become particularly popular among middle-class Chinese women. As the demand has overwhelmed the supply, these bags have become twice as expensive in China as in the United States. Naturally, many come to America to buy Coach. That's why there are often long lines in front of Coach Factory Outlet stores in the United States. Some such stores have to set a limit of ten bags for each customer to avoid running out of stock.

Envy-provoking marketing tactics can coerce consumers into purchasing items barely within their reach. In many countries, people of moderate means often sacrifice some essentials for a few luxury items. In nations that do not provide rigorous protection for brands, counterfeit goods such as watches, clothes, shoes, and handbags proliferate in order to satisfy such psychological needs. In China, for example, imitation and counterfeit commodities are so common that people often assure their peers that the products they are using are genuine, not fake.

A friend of mine, who used to run the product development and promotion unit in a Shanghai firm, once confided to me that the key to selling an expensive, nonessential product is to create a cultural pressure for consumption—that is, to make people who don't own it envy those who do. To be effective, advertising has to make average consumers feel inferior if they live without the product being promoted. Such pervasive commercial campaigns poke deep into the vulnerability of human nature, making us feel compelled to buy. Similar marketing ruses are used worldwide in TV commercials, fashion shows, store displays, magazines, and Internet ads. The two most successful advertising slogans today—"A Diamond Is Forever" and "Just Do It"—represent the triumph of this tactic. While some American young men in inner city slums trade their hard-earned money for the "right stuff"—cell phones and other expensive gadgets—to impress others in the neighborhood, horrifyingly, in 2011, a poor Chinese ninth grader by the name of Zhen allegedly sold one of his kidneys to buy an iPad.

As a result of envy-provoking marketing, conspicuous consumption has become widespread in many nations around the world. And it is highly contagious. In recent decades, the prevalence of consumerism has driven American consumer spending to a perilous level and left many Americans with meager savings and mounting debt. Writing in the *New York Times* on March 17, 2005, at the crest of the housing bubble, economist Robert Frank compared bidding wars over houses in top school districts to military arms races. As a consequence of competitive consumption, he noted that Americans save so little that nearly "a fifth of American adults have net worth of zero or less."[35] The economically savvy, on the contrary, do just the opposite. Warren Buffet, when asked why he kept some money tucked away, responded that it made him sleep well no matter what might happen in the financial markets tomorrow. If the average American thought like this, would our economy still swing as wildly as it does now? Driven by consumerism and sustained by envy,[36] "the savings shortfall," Frank warned, "threatens not just those who face retrenchment in retirement living standards, but also the country's economic prosperity."

Although envy can be benign, more often than not it fuels malicious acts. In the workplace, envy can lead to counterproductive behavior, low job satisfaction, and a low rate of employee retention, as those with high self-esteem tend to quit.[37] A study of 102 business units in forty-one British and American corporations demonstrates that the larger the pay gap between top managers and low-level employees, the lower the quality of their work product.[38] Likewise, the professional basketball teams with the greatest pay differentials among players tend to sink to the bottom in team performance and advertising revenue.[39] One may wonder whether some who occupy the lower rungs in these organizations may be doing something beyond mere complaining. These, together with many similar examples, illustrate that envy—while it may stimulate the competition to jack up performance—is often accompanied by significant social costs that result in even larger economic burdens.[40] Hence, envy is somewhat like fire: if you don't know how to put it out, it's better not to play with it.

Malicious envy often leads to lose-lose situations. Sometimes, the envier would rather go down with the target of his or her envy than allow the target to succeed. The downfall of the envied can make the envier happy. In Spinoza's words, such envy is "hatred in so far as it affects a man so that he is sad at the good fortune of another person and is glad when any evil happens to him."[41] This pleasurable feeling at the misfortune or downfall of another person is an emotion known as *Schadenfreude* in German, *skadefryd* in some Nordic languages, and *Xing Zai Le Huo* in Chinese. Oddly, English has no common word for it,[42] despite the fact that this emotion is not at all rare.

Here, Tiger Woods's scandal can be quite telling. While his imprudent lifestyle disgusted many Americans, it served as a source of Schadenfreude for others. Schadenfreude is also behind the widespread disdain for many celebrities—Britney Spears, Paris Hilton, Michael Vick, Lindsay Lohan, and Charlie Sheen, to name just a few—who have fallen after flaunting their fame as sports or entertainment stars.

To explore how common Schadenfreude is, a team of psychologists let university students watch a videotaped interview of either an average student or an ambitious premed student. Afterward, the researchers revealed that the student in the video was arrested for stealing a controlled substance from a research facility. Not surprisingly, the subjects who watched the video of the premed student demonstrated more Schadenfreude than those who watched the tape of the average student.[43] If we suspect that Schadenfreude in this case is simply a vindictive reaction that justice was served, our outlook on human nature may be a little too rosy. A follow-up study showed that the ruin of an enviable but innocent person for a reason totally irrelevant to justice can also elicit Schadenfreude.[44] An interesting question, then, is how Schadenfreude works in the brain.

A group of researchers at University College London have found that, at least for men, the brain areas related to pain are activated when seeing a friendly person experiencing pain. However, the brain areas related to reward, which are, interestingly, also linked to the perception of fairness and desire for revenge, light up when seeing a competitor receive the same pain.[45] This finding suggests that other people's pain can be perceived through two different neural pathways: one leading to sympathy and the other to pleasure.

Malicious envy prompts some people to take action, ranging from refusal to help the targets of their envy to committing crimes against them. One notorious example is the Tonya Harding saga. Harding, a good skater, won the Women's Singles in the US Figure Skating Championships in 1991. Even so, she was overshadowed by the better-looking, more skillful Nancy Kerrigan, who regularly defeated Harding in most ice-skating events. During the 1994 US Figure Skating Championships, Harding conspired with her ex-husband and boyfriend to injure Kerrigan before the competition. As expected, Harding, with her archrival sidelined, won the night—until the plot came to light. Harding was soon disgraced, though she was able to escape criminal charges through a plea bargain. She later made a name for herself in women's boxing. This later career could not have better suited to the tastes of the American public: to see her being knocked down and, better yet, out.

Harding's spiteful crime speaks to one fact: when critical resources are limited—the championship in this case—competition among peers can become a fierce zero-sum game. Under these circumstances, envy can easily become malicious, for few alternatives are available to the envier. Such situations can occur in what seems like the least likely group of people: nuns living in a convent. Driven by envy, there could be "extreme violence in certain communities" including "confinement, punishments, cruelty, deprivation of care, emotional abuse, poisoning, accusations of witchcraft."[46] Though nuns ideally dedicate their lives to God, a stoic lifestyle in a closeted microcosm apparently enables the all-too-common human emotion of envy to take the upper hand.

From these dramatic cases we can infer that in societies where basic resources are limited, with few avenues open for economic and social advancement, malicious envy will emerge or even prevail. Situated in just such a society, Jane was envied for her efficiency in farm work, even in a poverty-stricken collective community. Furthermore, she was simply too gifted, smart, and outstanding when compared with everyone else. She was perceived to be the person most likely to break away from the pack. She was the nail sticking out, the bird poking up its head, or, as Australians say, the tall poppy that needs to be clipped.

We have already learned much about enviers' motivations and actions. How then do the envied feel and react? To understand the mentality of the envied, two psychologists, Gerrod Parrott and Patricia Mosquera, led a cross-cultural study in Spain, the Netherlands, and North America. The researchers asked college students how they would feel if they won a prestigious internship in competition with a fellow student. The participants—bubbly, confident, and high in self-esteem in their roles as the winners—worried about becoming the targets of envy on the part of the losers. They felt sympathetic toward the outcompeted and guilty about their own success. To "make up" with the losing students—probably also to ameliorate their own "guilty" feeling—they would make friendly overtures, send compliments, give gifts, or provide a nice dinner to the defeated. (Does this sound familiar?) Interestingly, in North America, where personal success is valued the highest among the three societies, participants were also most concerned at the prospect of becoming targets of envy.[47] Individualism appears to compromise psychological well-being.

A suite of methods have subsequently emerged for the envied to avoid being the proverbial bird that pokes up its head. One way is to conceal items—wealth, jewels, even good food—that could draw unwanted attention and cause envy. Indeed, in Jane's poor community, people hid conspicuous advantages to avoid being envied. For instance, they ate anything considered "luxurious"—meat, fruit, even candies—in secret to avoid the glares from red eyes.

Many cultures have evolved fascinating tactics to address these concerns. In some Arab nations where boys are highly preferred over girls, newborn boys are often masqueraded in girls' clothes until the child's sex—at about five years old—is too obvious to conceal.[48] In Germany, where Mercedes are commonly used as taxis, some rich people who own top-of-the-line models chip off the shiny chrome series numbers to make their luxurious cars look plain. (Expensive models without series numbers can even be specially ordered in Germany.) Others park their Porsches in garages away from their homes, hiding the evidence that might enable their neighbors to notice their wealth.

Other common tactics for the envied include lowering one's profile or helping the less advantaged with gifts or donations. Sometimes the envied will even make themselves appear mediocre by deliberately reducing their effort and performance.[49] Warren Buffett, for instance, actively participates in charities and lives in a house far more modest than his wealth would allow. By the same token, Barry Bonds, the baseball player who holds the Major League Baseball record of most homeruns (762), had a reputation of being aloof with media and teammates, presumably to reduce the discomfort of being envied for his lucrative contracts and endorsement deals.[50]

The fear of being victimized by envy speaks to another paradox: envy can go beyond the border defined by societal norms of fairness. In fact, as long as there are gaps in status and wealth, regardless of how they are acquired, envy results and intensifies as the gap widens. (Remember how dopamine neurons work in the midbrain, discussed in chapters 2 and 3?) Therefore, the rich can become targets of envy, even if their fortunes have been legitimately acquired.[51] They are, as the popular slogan in the Occupy Wall Street (OWS) movements goes, the "1%" by default. Such sentiment can be found in the Christian Bible in the words of Jesus: "It is easier for a camel to go through the eye of a needle than for a rich man to enter the kingdom of God."[52] In 2003, for example, Dick Grasso, former chairman and CEO of the New York Stock Exchange, was forced to resign due to the public outcry over his $140 million compensation package. He once defended his excessive pay on CNBC by arguing that the company's board of directors made the decision based on how much he was worth. He was clearly arguing for his merit, in addition to the legality of his contract. In the eyes of the American public, however, his pay should not have been *that* much, regardless of his "worth." In this case, we can see the distinction between fairness in procedure and fairness in outcome that we encountered in previous chapters.

Here again, social comparison, underpinned by perceptions of relative payoff, plays a major role in public affairs. People's interest in the lives of the rich and famous is commonly an amalgam of awe and envy. With information regarding public figures constantly in front of our eyes—on television, in magazines, and especially over the Internet—the experiences of personalities such as sports stars and entertainment icons seem part of our everyday lives. Since the lavish lifestyles of superstars are well beyond our reach, envy of them can be more debilitating than inspiring, as benign envy occurs primarily with attainable goals.[53] It is almost inevitable that celebrities and wealthy people should become a common source of such envy. As a case in point, the former tallest NBA star, Yao Ming, donated a half million dollars to charity to help victims of the earthquake in Wenchuan, China, in 2008. Was he generous? Not enough, according to some critics, who accused him of being stingy. His annual income was over ten million dollars.

Bill Gates, the richest entrepreneur in the world for over a decade before 2010, supplies a more dramatic story. In the 1990s, he was broadly recognized as a self-centered billionaire and was a popular target of eggs, tomatoes, and pies in public, on both sides of the Atlantic. Since he set up the Bill and Melinda Gates Foundation in 2000 and announced he would donate most of his personal fortune to charity, his public image has undergone a radical makeover almost overnight—from arrogant, selfish billionaire to bighearted philanthropist. He has since been almost entirely immune from smearing attacks. The making and remaking of Bill Gates's public image show spectacularly the power of envy.

On the subject of coping with envy and being envied, Confucian scholars have offered plenty of wisdom. "Supreme indeed is the Mean as a moral virtue," claims Confucius.[54] His intellectual descendent, Zi Si, elaborated upon the idea by writing an entire book, *The Golden Mean*, or *Zhongyong* in Chinese. The text, allegedly based on Confucius's own words, was dedicated to the moral issue of fairness in society. The book advocates for three principles in human relationships: fairness, harmony, and practicality, in that order. The book says, in essence, that one must first hold a moral principle that is fair and just with no bias. In dealing with human affairs, one should never under- or overdo what is deemed appropriate. In handling conflicts, one should tolerate diversity in opinion and focus on compromise, seeking the middle ground. Finally, one should strive to be reasonable, flexible, and open-minded, adapting oneself to society.

Here, the word "Mean" or "Golden Mean" can be misleading. Confucius does not speak for mediocrity. On the contrary, he is highly critical of those who lack social ambitions and moral principles. "The village worthy is the ruin of virtue," he says. "[The gentleman] dislikes those who proclaim the evil in others. He dislikes those who, being in inferior positions, slander their superiors."[55] For Confucius, the importance of holding a solid moral ground is paramount. "The Three Armies can be deprived of their commanding officer," he declares, in perhaps the most powerful words in his moral teachings, "but even a common man cannot be deprived of his purpose."[56] "Purpose" here means moral commitment.

Confucius leaves us little doubt that he advocates excellence, social aspiration, and moral integrity. The reason he promotes the value of the Mean is that the Mean, as Confucius sees it, serves only as a means, not an end in itself. It is a tactical camouflage to avoid the slings of envy when one pursues one's ambitions in building a better society. In keeping with the Confucian tradition, the Chinese have developed an abundance of rules of thumb to survive social pressure against standing out: the knife with the sharpest blade should be hidden; a brilliant person should look as if he were stupid; a brave man behaves as if he were a coward. The message embedded in the time-tested Confucian wisdom is clear: unless the situation calls for you to step forward, you should stay humble and down-to-earth. We cannot help but be awed by the power of envy as a force of stabilizing selection, leveling socioeconomic status among people.

Sadly, Jane was too young to assimilate the Confucian wisdom and too naïve to grasp the gravity of becoming the unfortunate bird. Even if she had been sophisticated enough to realize her precarious position, it might still have been too early for her to succumb to the pressure from the noisy, red-eyed members of her commune. Jane thought that the worst she could suffer was to be constantly harassed or even socially banished. What she didn't know was that her commune was a microcosm shaped by a crushing society-wide force, a juggernaut that conspired to shatter her dreams. By defying it, Jane was assured a tragic ending. As a matter of fact, apart from China during the years of the Cultural Revolution, such a sweeping social force also hurt the United States and devastated the former Soviet Union in the twentieth century. Like Jane, many innocent people were robbed of their dreams and successful careers. Families were torn apart; lives were lost. This leviathan is known as anti-intellectualism. How it emerges, connects to our sense of fairness, and affects us even today is the focus of the next chapter.

Chapter 6

THE SOUL OF ANTI-INTELLECTUALISM

Another question that needs to be answered is why Jane was sent to a commune in the first place. This was, in fact, a gambit on Mao Zedong's political chessboard. The rise of anti-intellectualism in China during Mao's reign was intriguing. Governance by the learned had been among the core Confucian values in Chinese society since the second century BCE. For thirteen centuries from the Sui dynasty (581–618 CE) to the fall of the Qing dynasty in 1911, Confucian scholars had been recruited to government posts based on an open civil service examination system. In theory, this system allowed social advancement for virtually all people according to their levels of scholarly achievements.[1] With the constant infusion of fresh blood, this Confucian philosophy of governance by the wise seemed to work well. To a certain degree, it offset the widespread incompetence that resulted from a government rife with nepotism.

Mao, the founder of Communist China, was himself an intellectual. In fact, he was among the very few who had the privilege of some college education before the 1920s, when illiteracy in China topped 80 percent. Yet in the early years of his revolutionary career, Mao was overlooked, criticized, or sidelined by Communist intellectuals such as Chen Duxiu, Li Lisan, Zhou Enlai, Wang Jiaxiang, and Zhang Wentian. Several of Mao's rivals were university professors with advanced degrees from the West or the Soviet Union. In comparison, Mao, thought an elitist in education by any standard, was overshadowed by some intellectual heavyweights in the Communist Party. On several occasions, he was accused of being an inauthentic Communist for his unorthodox approach to guerrilla warfare. Only during the Long March—the Communists' year-long survival struggle after being routed from their base in Jiangxi by the Nationalist Army—was Mao able to seize an opportunity to secure his leadership in the Communist "Red Army." But a sense of intellectual insecurity, like a recurring nightmare, continued to haunt him. Even after consolidating his leadership in the Communist Party, Mao continued to be preoccupied by potential challenges to his authority from the intellectuals. After 1949, schools and colleges gradually returned to normal. Although the curricula were soaked in Communist ideologies, Mao was still concerned that people with formal education could pose a threat to his ruling.

Mao's prescience was confirmed in the early 1950s when some of his government officials who had limited education were put under fire by outspoken intellectuals for senseless policies. A common complaint in China at that time characterized the incompetence of local leaders as "non-experts leading the experts." Although Mao's level of education was matched by few, he still felt the pressure. Pompous, egotistic, and flamboyant, Mao deemed himself the greatest genius of all time. In his mind, he had no equal in any regard, not only in politics but also in a wide spectrum of intellectual pursuits—including poetry, literature, and calligraphy—that were valued and revered in traditional China. Comments and complaints from the learned appeared to open old wounds in Mao's deep-seated intellectual vulnerability. Sensitive and volatile, Mao took the legitimate criticisms as challenges to his policies, authority, and charisma.

In 1956, the central government launched the "Hundred Flowers" campaign, named after Mao's metaphorical instruction, "Let a hundred flowers bloom; let a hundred schools of thought contend." Mao at first encouraged people to speak out and criticize the government. But when some bold intellectuals went so far as to question the legitimacy of the Communist Party's power monopoly, Mao was shaken, believing such critics to be "the poisonous weeds" that must be rooted out. He made his move in 1957, staging a replica of Stalin's Purge—the Anti-Rightist Movement. Over half a million Chinese were designated as "anti-revolutionary rightists." They were removed from their positions and sent off to toil in factories and communes.[2] Hit hardest during the crackdown were organizations related to the arts, science, education, literature, the press, and other cultural institutions staffed by large numbers of intellectuals. As a result of this campaign, instead of blooming, the "Hundred Flowers" wilted.

Even so, stereotyping Mao as an evil dictator was an oversimplification that overlooked his vaulting ambition and profound personality. With the advantage of hindsight, Mao's greatest tragedy lay in his often competing cross purposes—on one hand, a drive to build a strong, unified nation, and on the other, a desire to consolidate his own power. Sadly, he succumbed to human weakness, letting his obsession for power run free and losing sight of his grand political vision. Such conflict of purpose could not be better illustrated than by his deep suspicion and distrust of intellectuals. After the Anti-Rightist Movement, Mao forged a sweeping policy to banish millions of intellectuals to the countryside for "reeducation." At least in part, the underlying motivation behind this was to consolidate his supreme leadership by eradicating potential political rivals.[3] The end result was tragic: much of China's brainpower was wasted, and the lives of a large number of Chinese intellectuals were ruined.

During the Cultural Revolution of 1966–1976, anti-intellectualism surged to an unprecedented level. Universities were closed and examinations were eliminated. Scientists, writers, journalists, and artists were sent to factories and farms to do manual work. Many middle and high schools were riddled with chaos and violence. In some cities, students and workers of different political cliques fought one another in the streets, under the name of protecting the Communist Party and Chairman Mao. In Chongqing and Shanghai, hundreds died during armed clashes and civil unrest.

A large number of officials, estimated at 60–70 percent of those staffing the central organs of the Communist Party, were removed from their posts.[4] Many top leaders were among the victims. Deng Xiaoping, for example, was repeatedly humiliated and beaten in public. His son, Deng Pufang, was permanently crippled after his failed attempt to escape torture by jumping out of a window at Peking University. The parents of several of today's Chinese leaders share Deng's mental and physical scars. Meanwhile, all middle and high school graduates in 1966, 1967, and 1968 were packed off to remote areas or the countryside for reeducation. Millions more educated young people would join them before Mao's death in 1976. This was why Jane, who barely qualified as an intellectual, was sent to labor in the rice paddies.

During the Cultural Revolution, Confucianism became an ideological enemy, poetry was replaced by revolutionary slogans, and literature was converted to propaganda. The arts became the voice of the Communist Party. Science was used to "prove" the correctness of Marxist-Leninist ideologies and Mao's thought. The performing arts and popular entertainments were dominated by the Eight Exemplary Dramas, including *The Red Lantern*, endorsed by Mao's wife, Jiang Qing.

The perverted upshot of the Cultural Revolution was the virtual eradication of China's rich and ancient cultural heritage. Anything related to literature, journalism, science, education, artistic expression, or several other aspects of culture was banished from public life. With this chilling attitude toward individual freedom and creativity, intellectuals were prevented from voicing opposing views. Peasants, soldiers, and factory workers, often semi-literate but loyal to Mao, were extolled as the paragons of society. Mediocrity was actively promoted and celebrated. Under this smothering and oppressive social environment, Jane's struggle to pursue her dreams was almost guaranteed to fail.

As a political maneuver to promote a cult of personality, anti-intellectualism was equally, if not more, disastrous in the Soviet Union under Stalin's reign. Stalin eliminated his more able and intellectual rivals such as Leon Trotsky, Nikolai Bukharin, and Sergei Kirov by banishing, murdering, or executing them, not to mention sending a vast number of former officials and intellectuals to the Gulag during the Great Purge of the 1930s. Lesser known, however, were the innumerable cases of the rank and file who used the occasion for their personal gain. Among them was the dramatic rise and fall of Trofim Lysenko, a self-made geneticist born to a peasant family with limited formal education.

In his quest for eminence, the mediocre yet ambitious Lysenko faced the obstacles of well-established Soviet geneticists, led by Nikolai Vavilov. Vavilov had an impressive track record of scientific accomplishments; he had collaborated in 1913 and 1914 with the renowned British biologist William Bateson, founder of genetics. During the period of 1929–1931, Vavilov was a member of the USSR Central Executive Committee, director of the Institute of Genetics, and the founder and president of the Lenin All-Union Academy of Agricultural Sciences (LAAAS).[5] He was a recipient of the prestigious Lenin Prize for his scientific achievements.

Lysenko, despite his ambition, was no match for Vavilov in science. His understanding of basic genetic concepts was often muddled, if not outright wrong.[6] To remove the major roadblock of Vavilov in his social ascendance, Lysenko could do little but resort to oblique methods, turning all his talent to political machination. He denounced the well-accepted Mendelian genetics supported by Vavilov and other mainstream geneticists as capitalist scholarship that was wrong for proletarian Soviet science. He denigrated the broadly accepted chromosome and gene theories by tying them to all the negative epithets he could imagine—"formalistic," "bourgeois," "idealistic," "metaphysical," "reactionary," and "barren."[7] He then deliberately elevated Ivan Michurin—an obscure native horticulturalist and a Bolshevik supporter—as a new icon of science. He promoted Michurin's Lamarckian theory of acquired inheritance—a mostly discredited evolutionary theory—as the orthodox dogma.[8] Meanwhile, Lysenko contrived a new "scientific" process called *vernalization*, claiming that soaking wheat seeds and storing them in snow could dramatically increase their productivity. To gather support for this theory, he created his own "scientific" journal, *Bulletin of Vernalization*, while at the same time pumping up media coverage with exaggerated results.

To suppress doubts about his vernalization method, Lysenko effectively transformed controversies regarding genetics into an emotionally charged political debate. By casting himself as the spokesperson of "proletarian" science, he led the crusade against the "bourgeois" science represented by his opponents. He claimed the clash of the two versions of science ultimately epitomized the class struggle in society at large. In 1935, in the presence of Stalin, Lysenko accused his opponents of being "no less dangerous, no less sworn enemies also in science." These "so-called scientists," Lysenko charged, "did their destructive business, both in the scientific world and out of it" before concluding, "a class enemy is always an enemy whether he is a scientist or not." This vehement diatribe won Stalin's applause, "Bravo, comrade Lysenko, bravo!" While his bluff was temporarily successful, Lysenko was aware of his own weakness. He once sheepishly admitted in private, "I have not studied Darwin properly."[9]

The questionable method of vernalization, promised to radically increase crop yields, made Lysenko a prominent figure in Soviet science. In reality, the practice of vernalization often lowered crop productivity, whereas the use of genetically selected breeds increased yields by 20–30 percent and, in some cases, by 30–50 percent.[10] But Lysenko rejected the proven genetic method in favor of his pseudoscience, on which he placed all his political wagers. Victories in politics, not in science, brought him many honors for his "scientific achievements." In 1927, he was fondly called "The barefoot Professor" by *Pravda*, the most authoritative government newspaper, for his peasant origins and lack of formal education.[11]

To silence his critics, Lysenko drew a large number of people into the debate over the two kinds of science, including natural scientists, social scientists, philosophers, school teachers, students, journalists, writers, and even farmers.[12] In the brouhaha of polemics, Lysenko diverted the issue from academia to the political arena. His politics-laden public campaigns won officials' approval and positive coverage by state media. Backed by Stalin in person, Lysenko was appointed the president of the LAAAS in 1935. He held this prestigious position for two decades without going through an election by the scientific community.[13]

Having consolidated his power, Lysenko began to cultivate a cult of personality for himself. He used his influence to make the government censor publications, ban theories that disagreed with his, cut articles and article sections unfavorable to his ideas from foreign scientific journals before they were allowed to be read, and remove his opponents from their positions, or, in many cases, imprison them, by tarring them "as the enemies of the people." He replaced these capable and well-educated scientists with incompetents, many of whom were his allies, cronies, and toadies.

Vavilov was Lysenko's best known victim. Accused of being a reactionary, spy, and saboteur, Vavilov was arrested on August 6, 1940, almost one year after he was elected president of the International Congress of Genetics in Edinburgh.[14] He was sentenced to be shot on July 9, 1941, but, for an unknown reason, his death sentence was converted to ten years in prison.[15] The Soviet authorities did not realize Vavilov's international fame and scientific achievements until months after he was elected a foreign member of the Royal Society of London in 1942. By that time, he was too ill to be saved. Vavilov died in prison on January 26, 1943. His premature death, historian Zhores Medvedev laments, was the "heaviest loss to Soviet science in the period of the personality cult."[16]

This academic cult, known as Lysenkoism, made the study of genetics anathema in the Soviet Union. As a result, genetic research in plants, cells, microbes, and related areas was banned. Many scientific laboratories, research institutes, and experimental stations were either closed or lost much of their state support. All materials related to chromosome and gene theories developed by Gregor Mendel and Thomas Hunt Morgan were removed from college curricula.[17] Lysenkoism brought disaster not only to a diverse range of biological and agricultural sciences but also to medicine and many other disciplines in natural and social sciences. Its influence persisted in the Soviet Union until 1964, when Lysenko was finally discredited and disgraced.

Because such a large number of talented scholars were removed, imprisoned, or even executed, the damage was monumental. Genetic research in the Soviet Union, after three decades of Lysenkoism, fell from a leading position to obscurity. In the West, meanwhile, genetic research bloomed with several major breakthroughs: the marriage between Mendelian genetics and Darwinian evolution gave birth to the sweeping success of the New Synthesis, the structure of DNA was discovered, and the deciphering of the genetic codes was well underway. While the West was poised to greet the new era of genetic engineering and biotechnology, in the Soviet Union, scientists were still debating whether the gene theory was correct. Unable to withstand this disaster, Soviet genetics didn't recover from the ruins of Lysenkoism until the dissolution of the Soviet Union in 1992. Ironically, the biggest beneficiary of Lysenko's anti-intellectualism—his effort in promoting mediocrity, his war on objective science, and his repression of scientific excellence—was the very archenemy the Soviet Union was so eager to crush: the United States.

Apart from Stalin's Soviet Union and Mao's China, oppression of intellectuals has occurred in many other nations: Italy under the control of the Fascists in the 1920s and 1930s, Argentina and Brazil in the 1960s, Iran after Ayatollah Khomeini's theocratic revolution in 1979. Interestingly, anti-intellectual sentiments are also pervasive in Great Britain, the United States, and other democratic nations.

Writing in 1963, historian Richard Hofstadter described anti-intellectualism as "a resentment and suspicion of the life of the mind and of those who are considered to represent it; and a disposition constantly to minimize the value of that life."[18] Anti-intellectualism, he added, is "a categorical folkish dislike of the educated classes and of anything respectable, established, pedigreed, or cultivated." He identified two entwined traditions of anti-intellectualism in US history: religious and populist movements. "Our anti-intellectualism," he wrote, "is older than our national identity, and has a long historical background."[19] In presidential elections, for instance, Thomas Jefferson was slandered for his intellectual achievements. Caricaturing Jefferson's interest in nature and inventions as "impaling butterflies and insects, and contriving turn-about chairs," William Loughton Smith, then a congressman from South Carolina pounded Jefferson:

> The characteristic traits of a philosopher, when he turns politician, are, timidity, whimsicalness, and a disposition to reason from certain principles, and not from the true nature of man; a proneness to predicate all his measures on certain abstract theories, formed in the recess of his cabinet, and not on the existing state of things and circumstances; an inertness of mind, as applied to governmental policy, a wavering of disposition when great and sudden emergencies demand promptness of decision and energy of action.[20]

Even more telling was what happened to John Quincy Adams, one of the best-educated men of his time, a great diplomat, and a farsighted abolitionist. During his reelection campaign in 1828, Adams was challenged, in an almost perverted way, from supporters of his opponent Andrew Jackson, a proslavery military man who backed the removal of Native Americans from federal lands.

Portraying Adams as a man utterly "self-indulgent and aristocratic," who "lived a life of luxury,"[21] Jackson appealed to vast numbers of ordinary white, male voters. This populist platform worked spectacularly. In the end, Jackson beat Adams in a landslide victory, winning every region but New England. Sickened by the vulgarity of superficial democracy, journalist H. L. Mencken wrote cynically a century later:

> All the odds are on the man who is, intrinsically, the most devious and mediocre—the man who can most easily adeptly disperse the notion that his mind is a virtual vacuum. The Presidency tends, year by year, to go to such men. As democracy is perfected, the office represents, more and more closely, the inner soul of the people. We move toward a lofty ideal. On some great and glorious day the plain folks of the land will reach their heart's desire at last, and the White House will be adorned by a downright moron.[22]

Why does anti-intellectualism occur in societies regardless of their political systems? It is understandable that, as pointed out by Hofstadter, "intellect is always potentially threatening to any institutional apparatus or to fixed centers of power."[23] In other words, intellectuals, for their pursuit of knowledge and reason, are often seen as a potential challenge to existing norms and established authorities. This perceived threat motivates some government and religious leaders to suppress intellectuals. It explains why Mao sent millions of learned Chinese, like Jane, to labor in factories and farms for reeducation, and why Stalin banished Soviet intellectuals en masse to the Gulag.

Since those in control benefit from protecting their status, prestige, and privileges, top-down anti-intellectualism makes sense. How can anti-intellectualism also become a bottom-up social force emanating from ordinary people? It's clear that intellectuals are commonly viewed as a privileged minority, enjoying certain social and economic advantages. In many societies, knowledge and education are passages to prestigious professions and government positions. Accordingly, they can be perceived as a major cause of inequality for those who lack intellectual achievements or, more importantly, educational opportunities. What Jane suffered was precisely this bottom-up anti-intellectualism, compounded by the top-down variety, as a result of the transgression of being talented and educated.

In the United States, anti-intellectualism "made its way into our politics because it became associated with our passion for equality."[24] The motive behind historical anti-intellectualism in America was a hostile feeling toward people with knowledge, education, and intellectual achievement, who were perceived as having unfair advantages in obtaining successful careers and high social status. Today, a love-hate sentiment toward intellectuals is still palpable for those who lack these advantages, as Hofstadter keenly observes:

> No one questions the value of intelligence; as an abstract quality it is universally esteemed, and individuals who seem to have it in exceptional degree are highly regarded. The man of intelligence is always praised; the man of intellect is sometimes also praised, especially when it is believed that intellect involves intelligence, but he is also often looked upon with resentment or suspicion. It is he, and not the intelligent man, who may be called unreliable, superfluous, immoral, or subversive; sometimes he is even said to be, for all his intellect, unintelligent.[25]

Such mixed feelings about intelligence and knowledge, like those about beauty and wealth, arise from an intense conflict in our minds. Most people want to be smart—in fact, the smarter the better. We desire a first-class education for a good career. For decades, and increasingly so, Ivy League colleges, regardless of their hefty tuitions, have been flooded with applications from all over the world. For the same reason, medical and law schools are among the most competitive educational institutions in America. As industrial societies become increasingly complex, more and more of the most lucrative jobs require intellectual talents and advanced education. In the United States, since the 1970s, the ranks of CEOs have been dominated by those with graduate training experience. By 1990, wages were already significantly skewed in favor of the highly educated: attorneys, physicians, engineers, architects, college teachers, and scientists.[26]

With the advent of the Information Age, hi-tech companies mushroomed. To maintain a competitive edge, many of them offer enviable pay packages to attract highly skilled workers. As corporations such as Microsoft and Google prospect for and recruit talents all over the world, the societal desire to be smart, skilled, and knowledgeable is stronger than ever. Many parents resort to early educational programs to give their children a head start. As a result of this cutthroat intellectual competition, any suggestion that denies or mocks our intelligence, knowledge, skills, or talents is deemed an insult. In 1994, for instance, Richard Herrnstein and Charles Murray's book *The Bell Curve* did just that. It touched the nerve of the American public and stirred up a national outcry for supporting data, demonstrating that blacks lagged behind whites in the distribution of IQ scores.[27]

Unfortunately, intelligence may not be achievable by effort alone. Like many other human traits, such as facial features, metabolic rate, and sexual orientation, intelligence is a product of numerous entangled factors in addition to personal effort: genetic heritage, nutritional level, educational opportunity, economic condition, parental support, social environment. Some of these factors are manageable; others are beyond our control. As a result, intelligence, commonly measured—and often misunderstood—by IQ scores, shows a bell-curve distribution in any population. Only a small proportion of people are either gifted or challenged, whereas the majority of people are clumped around the average. For this very reason, we are often torn between two conflicting facts: intelligence is good for social advancement, yet not everyone is equally gifted. As a consequence, intelligence, like wealth and beauty, can be the object of both desire and hate.[28]

From an evolutionary perspective, bottom-up anti-intellectualism, as Jane experienced in her community, is a leveling social force employed against those who deviate from society's intellectual norm. Such stabilizing selection should work against both the intellectually advanced and those who are intellectually challenged, the minorities at the tails of the bell curve. Why, though, is there little overt negativity toward those at the lower end of the bell curve in American society?[29]

The answer is historical prejudice. IQ testing gained notoriety a century ago when it was used as a gatekeeper to exclude those considered inferior, such as Southern and Eastern Europeans, from entering the United States. One "study" at Ellis Island in 1917 showed that immigrants labeled "feebleminded" represented as much as 83 percent of Jews, 80 percent of Hungarians, 79 percent of Italians, and 87 percent of Russians.[30] Why? Because the vast majority of immigrants so labeled were unable to speak English and were illiterate to boot. How could they score high on IQ tests administered in English? The abuse of IQ tests culminated during the eugenics movement, which led to the forced sterilization of 64,000 people in the United States.[31] The Nazis used eugenics as a pretext to justify their racial and genocidal policies.

The infamous and misguided eugenics movement has made people in Western nations extremely sensitive to discriminating against the mentally challenged. Moreover, because the underprivileged minority is unlikely a threat to anyone's social and economic advantage, showing our sympathy and compassion for the weak risks little in the way of undermining our own interests. On the contrary, it can enhance our public image and reputation.

However, for the intellectually gifted and well educated, public sentiment turns on its head. Many even feel it's fair game to reduce the competitive advantage of the gifted. This explains why anti-intellectual prejudice is still politically viable in democratic nations today. Furthermore, portraying an opponent as an elitist or a celebrity has become an effective political gambit. Apparently, bottom-up anti-intellectualism is an outgrowth of our fairness instinct. In this sense, Mencken was more incisive than amusing when he compared democracy to a giant engine of public envy.

The consequences of anti-intellectualism in democratic nations are comparable, to a certain degree, with those of the "Hundred Flowers" movement. McCarthyism, in the decade between the late 1940s and the late 1950s, remains a sober reminder that society-wide oppression against (mainly) intellectuals can and has occurred in the United States, a country long touted as a bulwark of democracy, transparency, and tolerance. The persecution was perpetuated under trumped-up charges of subversion, disloyalty, and treason, accusations parallel to being fingered as "anti-revolution" in Communist nations.

Backed by both Congress and the executive branch, the McCarthyites targeted a large number of innocent Americans—especially in government, entertainment, universities, and labor unions. During the witch hunt for purported Communists and their sympathizers, many ordinary Americans participated in fanning public animosity and spying on their fellow citizens to uncover evidence of Communist ties. By 1957, when McCarthyism was in its death throes, 140 American Communist Party members had been charged, and ninety-three had been convicted for subversive or criminal activities, based on little more than accusations. Many more lost their jobs and were excluded from future employment in their fields. Their constitutional rights of free speech and assembly were trampled; their reputations shattered; their careers wrecked. Left-leaning organizations were spied on for their "un-American" activities. Many books were banned or burned in some libraries,[32] a scene that could have been out of Mao's Cultural Revolution. Among the blacklisted or victimized intellectuals were the composers Aaron Copland and Leonard Bernstein, the actor Charlie Chaplin, the authors Arthur Miller and W. E. B. Du Bois, the chemist and two-time Nobel Prize winner Linus Pauling, the physicist and father of the atomic bomb J. Robert Oppenheimer, and the then obscure theoretical physicist Qian Xuesen.

"McCarthyism did more damage to the Constitution than the American Communist Party ever did," concludes historian Ellen Schrecker.[33] Her strong words aside, the damage done by zealous McCarthyites was much worse, more pervasive, and farther reaching than we realize today. This cannot be better illustrated by the defection of Qian, a little-known episode but a major blow to US national security during the reign of McCarthyism.

Before being swept up in the mania of McCarthyism, Qian was already an accomplished young scientist at Caltech and a cofounder of the Jet Propulsion Laboratory, a leading research facility for the US missile and space programs. His PhD mentor, Theodore von Kármán, spoke of Qian as "an undisputed genius whose work was providing an enormous impetus to advances in high-speed aerodynamics and jet propulsion."[34]

In June of 1950, however, Qian failed to pass his security clearance. He was accused of being a Communist sympathizer and dismissed from military research programs. Without a job and under constant surveillance, Qian wanted to go back to his native country China. As he was privy to cutting-edge secrets in rocket technology, Qian's request was denied by the US government. Moreover, he was held in isolation in a US Navy facility on Terminal Island, near Los Angeles. After five years in detention, he was finally let go in exchange for several US pilots captured by the Chinese Army during the Korean War.

Once back in China, Qian was charged with the task of developing Chinese rockets and missiles. He soon founded the Institute of Mechanics for this ambitious program. Under his leadership, the first Chinese missile, Dongfeng, was successfully launched in 1964. The better-known Silkworm missile was developed a few years later. Improving on the technology used for Dongfeng, China has recently developed a missile specifically designed for destroying aircraft carriers, dubbed the "carrier killer" by US defense analysts. Since few nations maintain fleets of aircraft carriers, the target is obvious. The technologies Qian helped to advance have also been used in Chinese space and nuclear programs, making China a major competitor for the United States in these areas today. Qian was duly recognized as the "Father of Chinese Rocketry." Without him, it is unlikely China could have caught up with the United States so quickly in rocket technologies.

Among the few people who had foreseen the consequence of Qian's forced departure was Dan Kimball, then Undersecretary of the Navy. "It was the stupidest thing this country ever did." He was furious after failing to keep Qian in America. "He was no more a Communist than I was, and we forced him to go."[35] Kimball allegedly claimed that Qian was worth a military force of five divisions and he would rather kill Qian than let him go back to China. Yet how many McCarthyites were able to foresee the consequences of their witch hunt for the long-term security of the nation they claimed to defend?

In the heyday of McCarthyism, few could grasp the practical and moral fallout of anti-intellectualism. Even President Eisenhower seemed to enjoy mocking the learned. "By the way," he spoke to a Republican audience in 1954, "I heard a definition of an intellectual that I thought was very interesting: a man who takes more words than are necessary to tell more than he knows."[36] Note that this was at a time when college education was uncommon among Americans. Popular sentiment was against not only the elitists who attended Ivy League schools, but also toward college education in general. McCarthy himself claimed that "the [educated] bright young men who are born with silver spoons in their mouths are the ones who have been worst." Some even viewed American universities as "the training grounds for the barbarians of the future."[37] How similar such opinions were to Mao's view that the well-educated are vulnerable to the danger of the capitalist ideology!

Before the twentieth century in America, the anti-intellectual sentiment was notably strong among businessmen and farmers, typically self-made and without formal education. Art, literature, and science—even applied science—found few followers and seemed irrelevant to the farming and business communities. Condemning such sentiment as a "crass, self-defeating kind of pragmatism,"[38] Hofstadter, in disbelief, found a disturbing paragraph in a publication by the US Office of Education in 1956:

> A considerable number of children, estimated at about four million, deviate sufficiently from mental, physical, and behavioral norms to require special educational provision. Among them are the blind and the partially seeing, the deaf and the hard of hearing, the speech-defective, the crippled, the delicate, the epileptic, the mentally deficient, the socially maladjusted, *and the extraordinarily gifted* [emphasis in original].[39]

Bizarre though it may seem today, the quote nonetheless echoes an intense public negativity toward intellectuals shortly before the downfall of McCarthyism.

The wake-up call did not come until the successful launch of the Soviet Sputnik, the first manmade spacecraft, on October 4, 1957. Literally overnight, Americans realized "the national distaste for intellect appeared to be not just a disgrace but a hazard to survival."[40] School and college curricula, especially in science and technology, were soon revamped, enhanced, and emphasized. This pro-intellectual milieu paved the way for the 1960 election of John F. Kennedy, one of the best-educated presidents ever to serve. Despite a presidential tenure cut short by assassination, Kennedy proved his brilliance and competence in handling the most daunting challenge America has ever faced: the threat of a nuclear war during the Cuban Missile Crisis. Yet after a few decades of peace, anti-intellectualism has snuck back with a vengeance.

The current wave of anti-intellectualism in America started at the end of the Cold War, when many Americans sensed a respite after the breakup of its mighty archenemy, the Soviet Union. This sense of relief has perhaps distracted America's focus from maintaining its leading position in science and technology.

One victim of American anti-intellectualism is the public school system. In recent decades, its quality has been far below America's economic standing among developed nations. A comparison of seventy-four nations showed that for fifteen-year-old students, the United States was ranked seventeenth in reading, twenty-third in science, and thirty-first in mathematics in 2009.[41] However, it led the world in education expenditures ($10,995 per student in primary and secondary education in 2008), exceeding the average ($8,169) of the thirty-five industrial nations by 35 percent.[42] Although there are many reasons for the high cost and poor outcomes in public education in the United States, it is hard to ignore an overlooked culprit: a culture of anti-intellectualism, which has been a major roadblock on the highway to improved academic performance.[43] In fact, what languages other than American English are so rich in derogatory terms for those who are intellectually gifted or put effort into learning—bookworm, geek, nerd, smarty, or more bluntly, smarty-pants, among many others? One repercussion is an unsupportive or even hostile learning environment, discouraging children who might otherwise excel in school.[44]

For years, concerned citizens have been alarmed by the loss of academic competitiveness among American children, and politicians have been ranting that public schools are failing. Lack of funding for school programs and incentives for teacher performance are often fingered as the problems. While these may contribute, the fact that the cost of US education is the highest in the world argues against their role as the primary causes. When routine drills in baseball, football, or basketball are admired, yet comparable drills in math, science, and language skills are criticized as rote learning, we sense the influence of anti-intellectualism on American children.

Anti-intellectualism hurts America even more in the way it infiltrates our political process, contributing to incompetent leadership. Most people agree that democracy is a good thing. It levels the political playing field and opens avenues for social advancement. Yet democracy is also a populist political market, where our mixed feelings about intelligence and intellectuals are shown, especially in presidential elections. On one hand, most Americans prefer experts in the market for services. We want the best physicians when we are sick, the best lawyers when we are involved in disputes, the best teachers to educate our children, even the best plumbers when we have a leaky pipe. On the other hand, when we select our president, who will be entrusted to handle the most complex and consequential issues of the nation, many voters settle for a candidate with whom they would enjoy having a beer, or one who looks like their neighbor. We require professional certification for doctors, teachers, nurses, barbers, bus drivers, inspectors, and pest controllers. Yet when choosing our president, the only requirements are age, residency, and a birth certificate; there's not even a simple test for basic knowledge about the US Constitution.

If failure is a popular characterization of the George W. Bush presidency, it's an understatement for journalist Jacob Weisberg, whose assessment, despite being pertinent, biting, yet free of personal attacks, is much grimmer:

> George W. Bush's two terms in office have merely been lost years, a period of ineptitude, neglect, and falling behind on a variety of domestic and international problems . . . . Because of his decisions, America has squandered much of its global leadership role, making itself weaker diplomatically, militarily, and economically. After its victory in the Cold War, the United States lost respect, support, and influence. This may take a generation to fix, or it may never be fully fixed, in which case the second Bush presidency will mark the beginning of a long-term decline in American status.[45]

Such sober appraisal is far from being mere hindsight. During the campaign for his first term, Bush's shortcomings in knowledge, intellectual curiosity, and ability to tackle complex issues was already evident, if not yet alarming. He was disappointing even to religiously motivated voters: he showed little understanding of theology or philosophy but awkwardly blurred out "Christ" as his favorite philosopher. "[Bush is] . . . someone who does not comprehend his limits or his motives," writes Weisberg. "Being president was something beyond Bush's capacities in a way he didn't recognize." Why was he still elected, not once, but twice?[46] Did voters bear some responsibility as well? Weisberg says yes. The presidency "is something he should never have been given a chance to do, and I continue to fault those who gave him the opportunity to fail more than I fault him for trying.[47]

A solemn lesson from the Bush presidency is that populist democracy can succumb to anti-intellectualism. This cannot be better illustrated than by *The Weakest Link*, a game show that originated in Britain but has been popular in many countries for several years. In the United States, each show had six or eight contestants vying for a prize of a potentially large sum. The prize amount grew if the contestants, in rotation, correctly answered a chain of trivia questions. The vindictive part of the game came at the end of each round, when one person was democratically voted out by the other contestants. The process was repeated until there were only two survivors, who competed for all the accumulated money. The person who correctly answered more questions took the prize whereas the loser, like all others who had been voted out in previous rounds, was left with nothing.

After an exhaustive analysis, economist Steven Levitt found few signs of racial discrimination against blacks in the voting process. Although older and Hispanic contestants were slightly more likely to be voted out throughout the game, these discriminations were overshadowed by discrimination against knowledge. In early rounds, competent contestants were welcome because they increased the pot by correctly answering questions, whereas weak contestants who contributed the least were most likely to be voted out. This early voting pattern was turned on its head in later rounds, as able contestants were perceived as direct threats for the ultimate goal—the prize money. For this reason, they were increasingly more likely to be voted out.[48]

One can see how well the Confucian wisdom of the Golden Mean works here. The strongest contestant, under such a democratic system, should hunker down and hide her brilliance. She should pretend to be mediocre, barely surviving until the final round, when she can fully exercise her brainpower to defeat the other finalist and take the prize. Who can blame people for such Machiavellian maneuvers when selection against the most intelligent runs strong? In retrospect, Jane's life might have been better had she adopted this strategy in her farming community.

This seemingly intrinsic flaw in universal suffrage leads to a hard question: is populist democracy vulnerable? For over a century and a half, since the time of Alexis de Tocqueville, this question has been rarely addressed in the West. Yet democracy has its shortcomings. Who could have foreseen the Athenian civilization, as glorious and democratic as it was for its time, falling prey to the militant Spartans and later to Alexander's Macedonians? In the previous century, we have seen the collapse of the democratically elected Chinese parliament in the 1910s and the devouring of the young Weimar Republic by Hitler's Nazis in the 1930s. In modern times, democracies in many Asian, African, and Latin American nations have been plagued by corruption, instability, chaos, or military coups.

These cases should be more than enough to caution us against the blind faith that democracy will always prevail. One can certainly argue that most of these episodes of failure do not fit well with American democracy. Though legitimate, this argument misses the point. For any vibrant political system, democracy in particular, its vitality has to be drawn from an internal mechanism that allows the system to reinvigorate itself. One key element in the mechanism is a strong collective brainpower, to state the obvious. Without it, the competitive edge of a nation will erode, and democracy will weaken. Moreover, as robbing the wealthy and smearing the beautiful are not solutions to, respectively, economic and biological inequality, anti-intellectualism is morally objectionable and practically ineffectual in eradicating intellectual disparity.[49]

Today, globalization has taken us into a world where the balance of power among nations is shifting at an unprecedented pace. To gain a competitive edge, many Asian governments—in Taiwan, Singapore, and mainland China, for instance—are staffed with large numbers of intellectuals. It is a situation similar to the post-revolution United States, when most of the Founding Fathers—Jefferson, Adams, Madison, Franklin—were original thinkers with outstanding intellectual weight. With high rates of GDP growth for more than two decades, China, in particular, has greatly narrowed the economic gap with the United States. Furthermore, when America and Europe were humbled by the collapse of mortgage-backed securities, the Chinese economy, little burdened by the housing bubble in the West, continued to enjoy a remarkable rate of growth. As the whispering that *Pax Americana* is on the decline becomes louder, the Chinese system, which relies heavily on wise decisions by the government, poses a real challenge to the American system with regard to economic expansion and the development of national strength. While it is too early to declare that American individualism, the free market, and populist democracy are outdated and outdone, it is certainly self-defeating if knowledge, education, and intelligence remain a political liability.

In part 2 we have examined how our fairness instinct can lead to envy in general and anti-intellectualism in particular. Again, the underlying ultimate cause is evolution's invisible hand, through the mechanism of stabilizing selection. Although malicious actions that result from garden-variety envy—such as those taken by Tonya Harding—are rare, they can have severe consequences. In the next part, we will zero in on a mutually destructive force—spite—and how it can lead to wild justice through revenge and retaliation. If envy, in the form of anti-intellectualism, can weaken a country, wild justice by revenge and retaliation can result in mass violence, which can topple a regime.

Part 3

JUSTICE BY ANY MEANS

Besides envy, a more consequential and grave outcome stemming from our fairness instinct is spite, by which peers can go down together in order to cut down the relative advantage of their opponents. Driven by attempts to get even and fueled by the wild justice of revenge and retaliation, spite, despite its ruinous nature for both parties involved, can often spiral out of control, resulting in a vicious cycle of mutual destruction. Spite lies at the core of such violent actions as duels, feuds, terrorist attacks, violent revolutions, and international conflicts, so commonly seen across the world in history and today. While illustrating the progressively dire consequences as the desire for justice grows from localized violence to international terrorism, this section also shows how forgiveness and intervention from a third party—especially law enforcement—have evolved as ways to break the cycle of spite.

Chapter 7

MASSACRE IN A VILLAGE

Ten miles east of Taiyuan, the capital of Shanxi province, is Dayukou, a village of about three hundred households with one thousand three hundred people. Until October 26, 2001, Dayukou had been a peaceful terra incognita in North China. On that fateful night, the village made national news for the first time. In a rampage lasting three hours, a local peasant, armed with only a double-barreled hunting rifle, killed fourteen villagers and wounded three others in eleven households.[1] In a nation where private ownership of firearms is banned, gun violence of this magnitude was unprecedented.

Twenty-five minutes before midnight, a call came in to 110 (the Chinese equivalent of a US 911 call), detailing the massacre. The entire local police force was instantly mobilized, and an urgent request went out to the provincial government for reinforcements. Over a thousand policemen were assembled in a matter of a few hours, casting a large dragnet with thirty-seven checkpoints covering every road leading out of the area. Fully armed police squads were deployed to comb through all the possible hideouts around Dayukou.

In contrast to the overwhelming police response, the suspect, Hu Wenhai, a forty-seven-year-old peasant, and his two accomplices, Hu Qinghai and Liu Haiwang, appeared to have no sophisticated escape plan. The massive manhunt turned out to be anticlimactic. In less than twenty-four hours, all three were taken into custody. Hu Wenhai and Liu Haiwang were caught in two separate taxis stopped at checkpoints, and Hu Qinghai was found in his home. None of them resisted. Despite the extraordinary nature of the crime, the arrest was unremarkable.

If the capture by police was routine, the suspects' motives for the massacre were intriguing in a way that seemed to defy common sense. Contrary to the stereotype of a cold-blooded murderer, the middle-aged ringleader, Hu Wenhai, appeared meek. By all accounts, he had been a good citizen with no criminal record and, even more puzzling, his household had been honored by the government as a model of unshakeable adherence to law and order. Throughout village affairs, Hu was known for his willingness to stand up for what was right, earning him the respect of the community. How could this man plan and commit such an egregious crime?

The troubles that led Hu down this path began with an incident he was not directly involved in. Shanxi is to China as Newcastle is to England—famous for rich coal reserves. In keeping with the region's reputation, the village had several small coal mines. In the 1990s, the Chinese government began implementing a nationwide policy to privatize state- and community-owned assets. Factories, firms, mines, housing projects, and other collective enterprises were transferred to private hands in the form of contracts, ownerships, or rights of use. At Dayukou, the operation of the coal mines was contracted to a company run by village leaders. The business was lucrative at a time when little could quell the booming Chinese economy's thirst for energy. In only a few years, the "coal bosses" would rise to be among the richest of the rich in China.

Several years earlier, in the late 1990s, a manager of the mines discovered that over a period of three years the mining company had evaded 1.25 million yuan (~$170,000) in taxes and fees by underreporting profit and productivity. He reported his findings to the authorities. The company executives—essentially, the village leaders—considered whistleblowing like this a bold act of betrayal, unlikely to occur in the absence of an organized conspiracy against them. They believed Hu, outspoken as he was in the affairs of the village, must have masterminded the leak of their financial malfeasance.

On the evening of June 19, 1999, around 9 PM, Hu was outside, finishing the day's work at his orchard. Suddenly, he was ambushed and assaulted with shovels by his neighbors, the Gao brothers. Luckily, his own brother, Qinghai, was nearby. Qinghai rushed to his brother's defense and fought off the attackers. Hu was immediately sent to a local hospital; despite the viciousness of the unexpected attack, he was extremely lucky to escape serious injury. Although he received twenty-three stitches, none of the three deep shovel gashes on his head turned out to be mortal. In deciphering the motivation behind the crime, Hu linked the unprovoked attack to the leakage of the mining operation's financial skulduggery. As he was among the very few who had the courage to speak out, he concluded that the village leaders must have chosen to act in retaliation.

Sometime later, Hu's speculation was ineptly confirmed by the village leaders, who approached him and offered more than the average annual salary of the time, 23,000 yuan (~$3,000) as a private settlement. Short-tempered and strong-minded, Hu summarily rejected the proposal. The brush with death was too traumatic for his anger to be easily assuaged. Instead, he opted for revenge by petitioning higher officials, a grievance procedure commonly used by rural people against their local leaders.

Despite the seemingly benign nature of his action, Hu's petition could be deadly to his enemies. The Chinese government had a track record of imposing harsh punishments on economic criminals. For instance, further south along the coast in Zhejiang province, one salesman was put to death in the 1990s for illegally siphoning ten million yuan from trade with foreign companies. At the time, financial fraud exceeding a million yuan could easily result in long sentences behind bars, if not worse, for the culprits.

Hu was positive that the government would mete out justice—and serve it in a spectacular way. He began to press his petition in earnest. With help from friends, he dug up more dirty laundry in the form of some five million yuan (~$700,000) in unpaid taxes the mining company owed to the government. Hu was ecstatic. With so much money involved, it seemed likely the village leaders were doomed. Elated, he visited one household after another to convince the villagers they had a case. In the end, 121 people were bold enough to step forward and sign their names to the petition. Hu delivered it to authorities from the town, country, district, and all the way up to the provincial government. Awaiting him, however, was a catch-22: lower officials asked him to go higher, but higher officials sent him lower. For eight long months, Hu shuffled up and down the bureaucratic ladder to no avail. He grew tired, frustrated, and desperate.

It appears that the village leaders were keenly aware of the ongoing action against them. As might be expected, they brought their personal connections and influence with higher authorities to bear in order to block the progress of the petition. Meanwhile, the villagers who had lent their names to the process grew antsy, worrying that they, too, could become targets of retaliation from the village leaders. Hu felt trapped. On one hand, his credibility as a folk leader, his reputation among the villagers, and his personal vengeance were all on the line with the petition; there was no way he could walk it back. On the other, he had already exhausted all possible resources and avenues, yet there seemed no prospect that justice would be served. As time went on, a make-or-break scenario took hold of his mind. He was now the proverbial fish caught in the net: either the fish would die or the net would be broken. As the official channels ran out, Hu began to reach the conclusion that the only way to serve justice would be through his own hands.

"Vengeance," for sociologists Pietro Marongiu and Graeme Newman, "is basically motivated by a concern of equality, justice, and reciprocity."[2] Revenge provides a prototype for the core concept of retribution in today's judicial systems. It has been pervasive as a means of meting out justice in both Eastern and Western societies from time immemorial. In the Jewish tradition, for instance, retributive justice was woven around the talion principle, or *lex talionis*, better known as "eye for eye, tooth for tooth, hand for hand, foot for foot, burning for burning, wound for wound, stripe for stripe."[3] It mandates that a wrong must be "corrected" by a wrong of equal kind, in equal proportion.[4] This principle served as the backbone for the Code of Hammurabi, the world's oldest legal code (it dates back to the eighteenth century BCE), which insists on symmetry in punishing offenders in exactly the same way that they had injured their victim(s). Besides "an eye for an eye," we can see many other examples of a retributive justice system that was based on institutionalized revenge:

> If a builder builds a house for someone, and does not construct it properly, and the house which he built falls in and kills its owner, then the builder shall be put to death.

> If a man strikes a pregnant woman, thereby causing her to miscarry and die, the assailant's daughter shall be put to death.
> If anyone strikes the body of a man higher in rank than he, he shall receive sixty blows with an ox-whip in public.
> If a man has put out the eye of a free man, they shall put out his eye.
> If he breaks the bone of a [free] man, they shall break his bone.[5]

In part, Hu's thirst for revenge drew on old Chinese traditions of justice, where vengeance is the nucleus of the folk code of conduct. Furthermore, revenge has been glorified and romanticized by the subculture of martial arts. Legendary kung-fu masters are said to have their own rules and ways of life. They live for justice; revenge at all costs is often their only mission and ultimate duty. Many centuries before the man and the legend were born, the spirit of Robin Hood had already been invoked in chivalric Chinese folklore, where strong, righteous people vanquished local bullies, crooks, rascals, and villains with their own hands. *The Heroes of the Marshes*, mentioned in chapter 5, is a well-read classic novel about peasant rebels, who, like Robin Hood and his ragged but ever-merry companions, rise to slay local scoundrels and rogue officials under the name of divine justice in the South Song dynasty (1127–1279 CE). It is difficult to overemphasize the power of such books. For generations, they have voiced the traditional meanings of social justice and served as literary models for actions deemed righteous.

The romanticized lives of fictional folk heroes reflect reality. Blood revenge was in fact common in ancient China; for a long time, it was not even considered a crime. On the contrary, blood revenge was taken as a duty to restore the honor of those who were killed by their feudal enemies. One of the most influential Chinese classics, compiled before the third century BCE, *Li Ji*, or *The Book of Rites*, mandates that, "With the enemy who has slain his father, one should not live under the same heaven. With the enemy who has slain his brother, one should never have his sword to seek [to deal vengeance]. With the enemy who has slain his intimate friend, one should not live in the same state [without seeking to slay him]."[6]

Put simply, if a close relative or friend is killed, one way or another, revenge is a must. To comport with the traditional duty of vengeance, Chinese law as recently as the Ming (1368–1644 CE) and Qing (1616–1911 CE) dynasties did not punish sons or grandsons for killing those who killed their fathers or grandfathers. Under such codes of social conduct and legal environments, clans might feud for generations to settle a score. Though sometimes dynastic governments tried to stamp out bloodshed between feuding clans, at the local level the practice remained common—and honored—until the early twentieth century. As a result of cultural inertia, violent clashes between clans or families still occur, albeit much less often, in rural areas of China today.

In addition to the weight of this crushing historical burden, Hu was the victim of an unfortunate double standard in China's evolving legal culture. Ancient Chinese society was run mostly by moral conventions. Civil laws, in particular, were deplorably inadequate; conflicts were resolved primarily by customs. In recent decades, although many civil laws have been added to the legal system, they have yet to fully take root. As recently as the 1990s, many civil laws were rarely enforced or taken seriously throughout much of rural China. Such a status quo helped folk culture persist in these areas. Locals continued to adhere to their traditional codes of conduct, based on revenge and retaliation. Many clung to the folk principle of justice expressed in a well-known Chinese aphorism: "It's not too late for a gentleman to get his revenge a decade later." Unsurprisingly, such a code of honor seems to have held sway in Hu's mind, instigating his fatal course of action.

Though ruthless during the killing spree, Hu wasn't entirely free from guilt. Yet his mind was awash with the tradition of Chinese retributive justice: life for life. Knowing it unlikely that his behavior would be condoned, he planned his own death. He packed fifteen sticks of dynamite, five blasting caps, and two explosive devices into a canvas bag in case his getaway attempt failed.[7] But, for whatever reason, he had second thoughts when his taxi was stopped and searched at the checkpoint. A moment of dithering cost him the chance to set off the explosives. The police reacted quickly and subdued him. In less than five minutes, he was in custody.

After the killing rampage, Hu became possessed by a sense of heroism in fighting for a just cause. There was little drama during his police interrogation, and no need for coaxing, bluffing, or threats in order to extract information. Hu was frank and cooperative—to a degree that shocked the police.

"Do you know why you were arrested?"

"Yes! I killed a couple of people," Hu answered matter-of-factly. He was strangely calm, speaking as if it were an ordinary conversation.

"A couple of people? You killed fourteen!"

"Only fourteen?" Hu was surprised.

"How many? You say it."

"I remembered it was seventeen."

"Fourteen of them died."

"I didn't see anybody still alive then. I poked every one of them, and added a couple of shots to those who were not dead yet. Apparently, I missed a few."

"Do you know the consequence?"

"Yes!" Hu turned to the interrogator and grimaced. "I have to pay for it with my life."

"Do you regret it?"

"How can't I?" Hu raised his voice, with an air of puzzlement as though he was surprised by the question. But then his tone softened with a tinge of repentance. "Well, I shouldn't have killed the boy." He was talking about a child who was not on the list of his targets. "I only knew after you told me he was paying a visit there."

Hu paused for a moment. "What I also regret is that not all who should have been killed are dead."

This excerpt of Hu's admission contains scant evidence of any remorse. In fact, his scheme was envisioned and executed meticulously from the start. According to his confession, he planned to take action on the eve of the coming Chinese New Year, when most families would gather together, feasting, celebrating, and watching entertainment programs on TV. Three months before the intended date of execution, however, Hu encountered an unexpected glitch. While extracting evidence of financial fraud in the mining company from a "suspect," Hu killed the man when he tried to escape. This unforeseen turn of events forced Hu to carry out his plot urgently, three months earlier than planned.

Mounted on a motorcycle that October night, with a hunting gun strapped on his back, Hu swept through the homes of his targets. Household by household, he left behind a tortuous trail of bloodshed and violence. Since not all his enemies were home, he was unable to finish them all. His self-appointed mission was only partially accomplished.

When a reporter later asked Hu why he didn't spare the innocent children of his enemies, he replied that if he failed to eradicate them altogether, they would surely take revenge on his own young son in the future. No doubt, Hu anticipated continued bloodletting and justified his action according to the ancient Chinese code of conduct: "A son must take revenge on his father's killer."

Hu's candor—and bluntness—facilitated the investigative process. Before long, all the evidence was in. The case was seamless. But the circumstances of the situation led to an unavoidable question: would Hu be condemned to death?

Historically, the Chinese principle of "life for life" was unyielding in meting out retributive justice. But in this case, many Chinese had the gut feeling that there was some room to argue for an exception. First of all, Hu was attacked first and, by convention, deserved the right to fight back. Second, unlike a lawless desperado, he had first sought resolution through official channels. Only after these options had been exhausted did he take the case into his own hands. Third, because his victims intended to murder him and retaliate on other whistleblowers, Hu acted to protect himself, his family, and those who had signed the petition. Finally, most of the victims were actually de facto criminals. Beyond simple bullying, they had committed major economic crimes, punishable at the very least by long terms in prison. Consequentially, as the folk logic goes, Hu helped the government do what it should have been doing in the first place.

These reasons, beyond justifying Hu's potential exemption from the death penalty, might also be what made Hu waver at the time of his arrest. A moment of hesitation eliminated the option of suicide. But now, this fact could be bent in his favor: Hu could argue that he resisted the temptation to set off the explosives because it would likely have hurt the taxi driver and the policemen arresting him. And, according to Chinese policy, leniency can be granted to those who confess their crimes and cooperate in an investigation. This is at least a partial explanation of why he was shockingly frank and cooperative. Adding all the conditions, Hu had some chance of being spared the death penalty.

For millennia, the Chinese have revered those who banish village toughs and evildoers. In ancient China, fathers were encouraged to execute their criminal sons as the ultimate selfless deed for the good of society. In accordance with these cultural codes of conduct, Hu believed he should be recognized as a hero, since he was brave enough to stand up for the people against injustice. It appeared that Hu had a chance, albeit slim, to avoid capital punishment. All guilt aside, he was hopeful that his life might be saved.

Even more to Hu's favor was that China, under pressure from international human rights organizations, had grown increasingly reluctant to execute criminals. Although the government did not openly admit this, it had made some tacit concessions to criticisms from the West. Meanwhile, as its economy soared, China had the luxury of being more merciful, and the trend was obvious. Stealing a million yuan could be more than enough for an economic criminal to be condemned to death in the 1980s, but in the 1990s, ten million yuan might be needed to justify the same sentence. Today, only in special cases will an economic criminal be executed.[8] It was unfortunate for Hu to have missed this pattern in the first place, when he opted for revenge through the government. But now he might be blessed as a potential beneficiary of the growing leniency in the Chinese criminal justice system.

In retrospect, as he went down the road leading to the massacre, Hu's insistence on revenge rather than settlement was the turning point. It set the tone for, and was a major step closer to, the tragic finale. Yet the burning desire for revenge that Hu felt is not at all alien to people in the West. A *Newsweek* poll on March 11, 1985, showed that 71 percent of respondents felt it was justified to take the law into one's own hands, at least under certain conditions. Without knowing the gruesome outcome of Hu's quest for personal justice, who would consider his choice unjustified? Indeed, how many of us could write off some sort of personal revenge as a legitimate reaction if faced with the same situation?

Wild justice, also known as vigilante or frontier justice, is the impulsive pursuit of revenge and retaliation disconnected from legally sanctioned state authority. The term "Wild Justice" is taken from the titles of books by Susan Jacoby, and Marc Bekoff and Jessica Pierce. As will become clear in the next chapter, it is an evolved behavior seen in many social animals.[9]

Wild justice is commonplace throughout history in human societies across the globe. In Japan, as in China, personal revenge prevailed for centuries with a sense of militant romanticism. Samurai culture, the die-for-honor spirit of Bushido, was rife in medieval times. Warriors were revered for loyalty to their masters and glorified for bravery in violent confrontations. They were obliged to pursue vengeance at all costs against enemies when their masters were killed. If their own honor and reputation were compromised, they would commit suicide by seppuku—gory disembowelment by stabbing one's own abdomen with a katana, often publicly. Before the Meiji Restoration of the nineteenth century, the government would grant permission to those who wanted to personally hunt down and slay those who killed their fathers or brothers. Some spent decades in pursuit of revenge.[10]

In European culture, family vendettas come to life in Homer's epics and Shakespeare's finest plays—*Hamlet*, *Othello*, *Romeo and Juliet*—and in numerous fables, sagas, and legends. These enduring stories reenact life in traditional Western societies, where revenge is a recurring theme. In the words of author Susan Jacoby, "Revenge, like love and the acquisition of worldly goods, is one of the grand themes of western literature, a fountainhead of epic and drama. It appears in every guise known to man and woman: as comedy and tragedy; as a sickness of the soul and as emotional liberation; as disgrace and as honor; as an enemy of social order and a restorer of cosmic order; as mortal sin and saving grace; as destructive self-indulgence and as justice."[11]

Medieval England—the spiritual if not actual homeland of Wild Justice, embodied by Robin Hood—was plagued by ruthless vendettas. Feuding was epitomized by the family war between Earl Uhtred and Thurbrand the Hold. Starting in the early eleventh century, the vendetta between the two families lasted fifty-seven years, spanning four generations. Although many family members died on both sides, the blood feud didn't end until the last man of Uhtred's lineage was killed.[12] Despite royal attempts to curb feudal warfare, the folk code of blood for blood in England persisted until private vengeance was outlawed by William the Conqueror.[13] Though this reduced the incidence of family vendettas, they refused to disappear.

Broadly speaking, personal vengeance has served as a means for resolving conflicts related to perceived injustice throughout human history. It is easy to find instances of revenge and retaliation in our own society. Defense of one's honor is an element of both upscale communities and poverty-stricken city slums, from nobles and aristocrats to bandits and gangsters. Famous intellectuals can also become embroiled in duels: the Swedish astronomer Tycho Brahe, the French mathematician Évariste Galois, the Enlightenment writer Voltaire, the Russian poet Alexander Pushkin. Even Napoleon's nemesis, the Duke of Wellington—despite his stern opposition to personal justice—engaged in a duel when challenged by the Earl of Winchilsea in 1829.

For a time, dueling was quite popular in France, Britain, Ireland, Germany, Italy, Russia, and other European countries. During the reign of Henry IV (1589–1610), over four thousand French gentlemen lost their lives in duels—a formidable legion of the fearless who might have sacrificed their lives more profitably (at least for their countries) in battle.[14] Prior to the middle of the nineteenth century, North America wasn't spared the plague of dueling. Besides the fatal clash between Alexander Hamilton and Aaron Burr, "Henry Clay fought in one, and James Monroe thought the better of challenging John Adams," wrote psychologist Steven Pinker. "Andrew Jackson . . . carried bullets from so many duels that he claimed to 'rattle like a bag of marbles' when he walked. Even the Great Emancipator . . . Abraham Lincoln, accepted a challenge to fight a duel, though he set the conditions to ensure that it would not be consummated."[15]

History shows time and again that the get-even mentality is a common instigator of human conflicts. Recent surveys confirm that the desire for vengeance continues to motivate people around the world. Only a few decades ago, conservative estimates put the rate of revenge-motivated homicides at 8–9 percent in nations or regions as diverse as Australia, Ireland, and Hong Kong.[16] A close inspection of the statistics in the United States puts the rate at around 20 percent.[17] In fact, of sixty societies dominated by indigenous cultures around the world, all but three explicitly exemplify the desire for blood revenge.[18] Personal vengeance is apparently far more than a passing cultural peculiarity. As such, it's difficult to avoid the conjecture that the desire for blood vengeance is rooted deeply in our basic biological instincts.[19]

Revenge is motivated by exactly the same cognitive dissonance that defines injustice: the discrepancy between what we think ought to be and what is. Hu did not have to murder his enemies, as he said in his deposition. He was relatively well off, earning 40,000–50,000 yuan while the average in his village was around 10,000–20,000 yuan at the time. He could have simply shrugged his shoulders to those who were bullied by the village leaders. However, the confluence of his sense of fairness and justice, his cultural burden of chivalric honor, and the indifference, ineffectualness, and corruption of the state bureaucracy prodded him to take the most extreme action—to serve justice by himself.

Throughout the ongoing ordeal, Hu had two distinct exits. Internally, he could have forgiven his attackers and the perpetrators behind the events by accepting their offer for settlement. Externally, the government could have intervened. Had either occurred, the tragedy might not have come to fruition. Unfortunately, Hu gave up the first, which was precarious at best, and the government refused to grant him the second.

Hu's case was a judicial outlier; it was the worst gun violence in modern Chinese history. This fact alone would almost certainly condemn Hu to death. However, there were also plenty of reasons to spare Hu from capital punishment. For the Chinese public, the strongest was the consequence of his action—he had rid the community of several rogue officials, bullies, and criminals. Public sentiment was fueled, in part, by anger against the rampant corruption of government officials. Because of this, Hu's case evoked a great deal of public sympathy; some even hailed him as a folk hero.

Historically, public perception held sway in the application of criminal justice in China. If the public condemned a criminal, he or she would likely receive a much harsher sentence than if the public was more forgiving. Rape, for instance, was considered a grave crime, sternly condemned by the public, and, because of this, a convicted rapist was usually punished by a long term in prison or, in some cases, the death penalty. In contrast, a positive popular reaction might lead to a much lighter penalty.

During the trial, Hu, basked in his self-assured manliness, plainly acknowledged the killings. He defended the innocence of his two accomplices. Even after both conceded they had taken part in extracting information from their victims, Hu still claimed that he forced them to do it. In doing so, he demonstrated his ultimate loyalty to his friends (Qinghai was his brother as well), epitomizing another characteristic of traditional Chinese morality: gallantry and courage.

When asked what he would like to say in court, Hu stood straight with the air of a hero and delivered a lengthy statement with a resounding voice, "I was born in the New Society and grew up under the Red Flag. . . . I hoped I would become a righteous and kind person, a goal that I have strived to achieve. From a very young age, I have been brave and outspoken, daring to do whatever is right and good for people. I got along well with the kind yet powerless people in my village. Sometimes, I was the only voice for them.

"In recent years, a succession of village officials have embezzled public money, bribed authorities, and bullied the villagers. They have privately carved up and pocketed the four million yuan in taxes collected from the village coal mines and other local businesses. I, together with several other folks, have reported these cases to various levels of authorities during the last four years, but all have fallen to silence like a rock sunk to the bottom of the sea.

"Authorities from the Police, the Disciplinary Inspection, the Prosecutorial Departments of the District, City, and Provincial Governments have all treated us with indifference and contempt . . . . Where should we present our case? Who then are willing to serve justice for us? When I reported the case to the police, these civil servants were only busy driving their 300,000-plus-yuan cars around and had no time to deal with the case.[20] They even hooked up with the village cadres to bully the villagers . . . . I had no choice but to use violence against violence.

"In reality, I have made forty to fifty grand [yuan] a year. I could have completely turned a deaf ear to the grievances [of the village people}. However, I was unable to do so; my conscience told me that I should not do that; I just can't ignore the issue . . . ."

The audience in the court gave Hu's statement a loud round of applause. The presiding judge intervened to quiet them. But the cheers for Hu could not be stopped outside the court. Some regarded him as a national hero. Many urged the government to grant leniency. The Internet provided an unprecedented venue for the expression of a diverse range of opinions. In fact, the blogosphere was saturated with admiration for Hu.

Hu's own sense of heroism struck a chord with popular sentiment in China. Only a small minority of people accused him of being lawless, reckless, and cold-blooded. Even these critics conceded that Hu's original intentions were good. A large number of people were on his side. If the Chinese government did respond to public reactions according to tradition, Hu's life stood a chance of being saved.

Unfortunately for Hu, contrary to public will, the court did not buy it this time. It handed down the death penalty to Hu and his brother Qinghai on December 25, 2001. The second accomplice, Liu Haiwang, was sentenced to life in prison.

The disparity between reality and Hu's hope arose from his failure to sense the new pulse of the Chinese criminal justice system: the court had been increasingly independent in handling legal issues in recent years. At a time when China was trying everything possible to build a positive image in the world, the chance that such a well-publicized, high-profile case would be swayed by public opinion was slim.

Hu appealed to a high people's court, the highest court at the provincial level, for mercy, but his plea was quickly struck down. On January 25, 2002, Hu was slated to be executed by firing squad at 10:30 AM. But Hu, having long prepared for the worst, was fearless. That morning, though his hands were cuffed, he was still nodding to the audience like a hero, using body language to say farewell to the world. On his way out, he somehow got hold of a guard's hand and uttered, "I'm sorry I have to leave early." He spoke politely, without regret, as if he were departing from a friend's house where he had been kindly hosted.

While we can debate whether Hu deserved the death penalty, his case exposes a more serious problem. The government failed both Hu and the villagers on a massive scale by neglecting his original petition. It breached the sacred promise that only the state has the authority to mete out retributive justice. When this powerful muscle was paralyzed by bureaucracy, the state lost its credibility as a sacrosanct body, an impartial arbiter of social conflict. This loss of authority became apparent through the public approbation of Hu's behavior. But there was a more worrying ramification that resulted from the failures of the judicial system: it encouraged people to continue to seek wild justice through revenge and retaliation. "In a world of law," writes Jacoby, "the absence of just revenge poses as great a threat to both liberty and order as revenge gone wild."[21] The real tragedy lay in the perilous possibility that Hu's tragedy might not be the last of its kind. Indeed, there have been few signs that revenge-motivated homicide in China is on the decline.

From a broad perspective, our cultural institutions have radically altered the adaptive landscape of our past. Some of our instincts shaped in ancient environments are necessarily out of sync with the cultural rhythm of modern society. With the powerful state assuming the duty of serving retributive justice, our instinct for personal vengeance becomes all the more maladaptive—even while we still tend to instinctively fall back on it when circumstances demand. Hu was cursed because he was caught at the crossroad between ancient and modern ways and made a wrong turn. He lived in a time when the hold of the medieval code of conduct—gallantry, loyalty, revenge, and chivalry—still held sway, particularly in the life of rural villages. By contrast, the central government of China was striving for a civil society under the rule of law by shaking off the grip of ancient moral customs. Fooled by a quixotic belief in traditional folk values, he pinned his faith on revenge as a virtue, when, in modern reality, it was a crime. Ultimately, Hu was crucified for being an anachronism, caught between his genetic and cultural heritage and today's ever modernizing Chinese criminal justice system.

Hu's case is exceptional, but it is sadly just one of many that illustrate a crucial point: wild justice is still widespread, more common than we're aware of. Psychologist Michael McCullough provides a hodgepodge list of what can happen when people choose revenge and retaliation:

> arson, gossip, school bullying, urinating in the coffee maker in the break room at work, taking a long time to leave a parking space after someone has honked at you, road rage, World War I, World War II, workplace shootings, the bombing of a Tel Aviv pizza parlor, stealing stuff from work, giving away national secrets, the Hatfield-McCoy feud, the Alexander Hamilton–Aaron Burr duel, sports-related violence, voting against a colleague's promotion, vandalism, having an affair, shooting an unfaithful husband or wife, gang warfare, intentionally infecting someone with HIV, shoplifting, procrastinating, assassinations, and invading foreign nations.[22]

In some cases, like Hu's, revenge against society can be horrifically violent, moving from simple cases of personal revenge to include serial killing and domestic terrorism. Timothy McVeigh, the most notorious domestic terrorist ever produced on US soil, was driven by his own version of justice in response to government actions at Ruby Ridge, Idaho, and Waco, Texas. The Oklahoma City bombing was wild justice in its most extreme and deadly form.

The sheer normalcy of vengeance and its disturbing ability to intensify and spin out of control, as it has so often done, make it imperative for us to unravel the nature of the various forms of revenge and retaliation. Is there a common theme among them? Why does the impulse to take revenge, despite the potential of mutual destruction, have such a continuing hold on our behavior? How can we contain or overcome its powerful influence? These far-reaching questions are begging for answers.

Chapter 8

THE ORIGIN OF WILD JUSTICE

The tenacity of our desire for revenge signals something deeper that underlies our cultural norms. For decades, the mystery of vengeance has been a focus of persistent investigations from a wide range of scholars in multiple disciplines—biology, psychology, economics, anthropology, sociology, and political science. An early clue emerged in 1960 from a little-known study. It produced evidence of a puzzling phenomenon that seemed to go against the grain of conventional wisdom.

A group of researchers at Ohio State University asked college students to play a simple two-person, non-zero sum game.[1] In the game, two players who are separated by an opaque screen can push two buttons, one for "Cooperate" and the other for "Defect," with the following payoffs:[2]

Table 3. The payoff matrix of a Cooperation Game with absolute values

		Player 2	
		Cooperate	Defect
Player 1	Cooperate	4¢, 4¢	1¢, 3¢
	Defect	3¢, 1¢	0¢, 0¢

Modified from data presented in J. S. Minas, A. Scodel, D. Marlowe, and H. Rawson, "Some Descriptive Aspects of Two-Person Non-Zero-Sum Games, II," *Journal of Conflict Resolution* 4 (1960): 193–97.

Here, the paired numbers in each of the four cells are the payoffs for Player 1 and Player 2, respectively, corresponding to the strategies ("Cooperate" or "Defect") they choose. For instance, if both players choose "Cooperate," their payoffs are the same, 4¢ apiece, or (4¢, 4¢), as presented in table 3. If Player 1 chooses "Defect" and Player 2 chooses "Cooperate," the payoff is 3¢ for Player 1 and 1¢ for Player 2, or (3¢, 1¢), as presented in table 3.

Given the payoff matrix in table 3, what will the players do? The logical answer is that they should both cooperate because this will yield the highest profit for both (4¢ per round). Since mutual cooperation yields a better return than any other strategic combination for both players, it appears unbeatable.[3] The game is accordingly called a Cooperation Game, because the best strategy for both players is to cooperate. When real people played the game in the study, however, the result deviated far from the prediction: nearly half the players chose to defect! If evolution through natural selection has optimized our drive to maximize our own fitness, how can we fall so far outside the ballpark of rational behavior?

The answer lies in a hidden trick in our reasoning process. In this game, what we consider "rational" is viewed through the lens of absolute payoff. The real calculus used to gauge fitness, however, is often relative payoff. Say you are Player 1. While in an absolute sense you'd be better off by playing "Cooperate," you can't do better than your partner, Player 2. However, if you play "Defect" while your partner plays "Cooperate," you will be better off, gaining 3¢ versus 1¢. If your partner follows the same logic, mutual defection—the worst combination of all outcomes—prevails. This point is made clearer by converting absolute payoffs into relative payoffs to show the difference between the two players in each cell in table 3:[4]

Table 4. Payoff matrix with values

		Player 2	
		Cooperate	Defect
Player 1	Cooperate	0¢, 0¢	-2¢, 0¢
	Defect	0¢, -2¢	0¢, 0¢

Table 4, where the difference between the players is emphasized, demonstrates that the best possible relative payoff for you is the same for either strategy, but if you play "Cooperate," you risk being suckered when your partner plays "Defect": -2¢ for you and 0¢ for your partner. So, playing "Defect" guarantees the best possible relative outcome, regardless of what your partner does.[5] Since your partner is also a thinking *Homo sapiens*, you and your partner become locked in mutual defection. What begins as a Cooperation Game in table 3—where absolute payoff is what counts—degenerates into what I call a Spite Game in table 4, where it is the relative payoff that matters.[6] Worse than any zero-sum game, where there is *always* a winner and a loser, a Spite Game can result in two losers. Even if a winner does eventually recuperate from mutual destruction, both parties are worse off in the absolute sense for the time being. But why are we so prone to playing the Spite Game?

Zoologist David Barash, a pioneer in sociobiology, interprets it as a consequence of both players attempting to maximize the difference between them as they compete to gain the upper hand "in a small tribal band."[7] Therefore, it seems evolution hasn't created a major flaw in the human brain, after all. Instead, it operates in a manner that seems stunningly counterintuitive. More so than following the track of absolute fitness, evolution proceeds along the highway of relative fitness. In guiding our interactions with others, it is often the relative payoff that counts. Envy is a good example, discussed in substantial detail in part 2.

But relative payoff accounts for only half the players who choose to defect. Why do the other half still choose cooperation? The explanation again seems to arise from the familiar better-than-the-rest-of-the-pack effect. Barash points out that because our social psyche is tuned to handling affairs in small hunter-gatherer bands, a two-person game is a gross oversimplification of the conditions in which the human mind evolved. Even though participants in the experiment were restricted to considering a two-person game, how likely was it that their minds could ignore the existence of others in their social group? So absolute payoffs still matter because even the loser may do better than the rest of the pack. For any player, a large absolute payoff for cooperation also brings a large relative payoff against the rest of the pack, if not against the other player. Thus, absolute payoff, depending on its size, can still prompt two players to cooperate, a scenario similar to the high-stakes Ultimatum Game discussed in chapter 3.

Here, if we raise the rewards for mutual cooperation from (4¢, 4¢) to ($4, $4), ($400, $400), or even higher, can we expect the players to become more cooperative? No one has done this experiment. But human life and human history are replete with examples showing that large stakes (that is, absolute payoffs) favor cooperation. The Western Front in France and Belgium during World War I offers a case in point. Instead of fighting, British and German soldiers in some divisions refrained from killing each other as the combat wore on. A live-and-let-live system emerged along several sections of the five-hundred-mile-long front. In some cases, soldiers could stroll between the trenches with little chance of being shot. They intentionally missed each other's targets, called out for truces during holidays, and apologized to the enemy for casualties caused by their artillery units, which were stationed far away. The giant payoff of saving lives triggered the spontaneous emergence of mutual cooperation, even under these least likely of circumstances.[8]

The majority of our more mundane everyday cooperation, however, confers only a minute advantage, often barely noticeable. Good relationships at home and work mostly reward us with smiles, comfort, or conveniences; they rarely involve millions of dollars or life and death. Therefore, relative payoffs tend to prevail in these small-stake situations, making it easier for our social interactions to slip into mutual defection. This may explain why cooperation is often tenuous, easily succumbing to the siren call of defection. The intrinsic fragility of cooperation offers a rationale for Nietzsche's pessimism, "On the average, a small dose of aggression, malice, or insinuation certainly suffices to drive the blood into the eyes—and the fairness out of the eyes—of even the most upright people."[9] No wonder even a dedicated friendship, a romantic love affair, or a happy marriage is so hard to forge, yet often so easy to ruin. Sometimes it takes only a minor offense or a misunderstanding to turn a good relationship upside down. One sentence uttered in anger can nullify years of nurturing, loyalty, and dedication.

Mutual defection, destructive to both partners involved, is known as spite. Guided by evolution's invisible hand, spite always lurks under the surface and prevails when conditions are right. In addition to complicating human relationships or instigating price wars between competing businesses, spite plays a role in the behavior of a number of organisms, many of which have no brain. Bacteria, for example, can manufacture toxins that kill their conspecific neighbors. We've learned to exploit many such chemicals as antibiotics. Meanwhile, sticklebacks raid nests to munch hidden eggs of their own species, even when food is abundant or the eggs of a sibling species are so readily available that they're virtually begging to be eaten. Swallows work hard to defend large territories—not for the welfare of their own chicks, but to reduce the reproductive chances of their competitors. Monkeys often harass couples in their cohort who are mating.[10] This may explain why humans have evolved the habit of making love in private. Cases of spite in the animal kingdom go on and on. Biologists have only uncovered the tip of an immense iceberg, with endless variations of spiteful behavior concealed beneath the surface.

If there is anything special about human behavior, it's our ability to reject and rise above our biologically ordained fate. This tendency can assert itself when we face the evolutionary imperative of spite: though we are not immune from it, we are able to muster the courage to understand it and, ultimately, attempt to short-circuit its influence on our behavior. One of the scholars leading the effort to understand spite, and our ability to circumvent it, is Robert Axelrod, a University of Michigan political scientist. In the late 1970s, he did something unusual for a researcher in his specialty at that time: play computer games. He invited professional game theorists in mathematics and the social sciences to participate—by submitting computer programs—in a game called the "Iterated Prisoner's Dilemma," with the following payoffs:[11]

Table 5. Payoff matrix used in Axelrod's study

		Player 2	
		Cooperate	Defect
Player 1	Cooperate	3, 3	0, 5
	Defect	5, 0	1, 1

Modified from data presented in R. M. Axelrod, *The Evolution of Cooperation* (New York: Basic Books, 1984), p. 8.

(Prisoner's Dilemma is a game whose payoffs bear some similarity with our Spite Game, with mutual defection as the unbeatable strategy). After receiving fourteen computer programs, each containing an explicit strategy, Axelrod let the strategies play against one another in a round-robin fashion: each entry played every other entry in addition to itself and a blank control called RANDOM, a tactless strategy that cooperates or defects randomly. When two programs met, they played each other two hundred times in the same game—thus the name "iterated." When the entire round-robin tournament of 120 such pair-wise games was completed, the strategy that scored the highest total won. It turned out that a strategy by the name of TIT-FOR-TAT, or TFT for short, submitted by game theorist Anatol Rapoport, a professor at the University of Toronto, was the winner.

TFT is so simple that it has only four lines of code in Fortran,[12] compared with seventy-seven lines for another strategy—clearly much more complex—submitted by a person whose name was never released. Simple as well as guileless, TFT starts with cooperation, and, after that, it always plays whatever its opponent played in the previous move—cooperate if its opponent cooperated, and defect if its opponent defected. Thus, TFT is a reflexive, mechanical reaction with no memory beyond the previous move. Axelrod himself was amazed at how such a simple copycat strategy, with little hindsight or foresight, could emerge the winner.

Before Axelrod could declare TFT unbeatable, there was an obvious problem. Although all the submitted programs were from leading researchers, fourteen entries were too few to crown TFT the champion. He needed to explore the issue further before he could be assured this strategy was invincible. One practical option was to open the tournament to additional challengers, inviting entries from the best minds in the world. So he did. Moreover, he tempted the contestants with a report on the outcome of the first round, including detailed comments on the pros and cons of each strategy, with the hope that this would elicit more successful strategies for the competition.

The trick worked. Axelrod received sixty-two entries from six countries. Among the competitors were biologists, physicists, computer scientists, and computer hobbyists, in addition to several of the players from the previous round. Some strategies were so intricately contrived that they ran well over a hundred lines of code. Several of them were variants of TFT with the implicit intention of outsmarting the original. Yet despite the high level of competition, the sophistication of the strategies, and the clear goal of defeating TFT, the simple TIT-FOR-TAT strategy confidently submitted again by Rapoport still emerged the overall winner.

Close examination shows that by mimicking its opponent's strategy, TFT won not by beating other strategies—instead, it broke even with its opponents in each and every dyadic match. More specifically, as a copycat, TFT gained an advantage by scoring high whenever its opponents scored high through cooperation. Ironically, by sticking to the rule of never being the first to defect, TFT also never won even a single game. However, it outperformed every other strategy when the scores were tallied at the end. So the Tao of winning for TFT is to be an unmoving mover: no matter what you do, my response is always the same.

Is TFT, as simple as it is, the best possible strategy? Despite this confirmation, Axelrod appeared unwilling to give a definitive yes, and the question remained open. Soon, behavioral biologists entered the fray. One of the most interesting discoveries was made in small fishes. Guppies, for instance, are food for many larger fish. When venturing out into open water from their safe havens in weeds, they often gang up for the mutual benefit of safety in numbers. Indeed, even if only two guppies travel together, the risk is cut in half for both, compared with going alone. But there is a rub: a cunning guppy can "promise" to join the adventure, yet when seeing its comrade go, stay back and watch the lead fish test the water alone. In this case, the solo lead fish is at greater risk than the trailing one. What should the lead fish do to prevent such an unfair joint venture? You may have already guessed correctly—expose the trailing fish by swimming back. In fact, according to the tally by Lee Alan Dugatkin, the lead fish is twice as likely to turn back as the trailing fish. Apparently, by doing so the lead fish ensures it will not go alone. Dugatkin interprets the frequent turning back by the lead fish as a retaliatory measure, an element of TFT, aimed at enforcing the cooperation of the trailing fish in this simple survival game. Further studies in guppies and sticklebacks confirm the use of TFT for this purpose.[13] Thus, TFT isn't successful only in computer simulations; it's an effective strategy for real-world organisms as well.

How reliable is TFT? Axelrod explored this issue by designing a survival-of-the-fittest scramble among all these strategies, again using computer simulation. He first threw together all the sixty-two strategies, plus RANDOM for control, and let jungle law decide their fates. After the initial melee, he allowed each strategy to "reproduce offspring"—copy itself, like real organisms—in proportion to its score. Then he ran the melee again. And so on, until a thousand rounds were completed. He found that TFT again came out on top, rising from one out of sixty-three entries in the beginning (~1 percent) to over 14 percent of the total at the end. He also discovered that the fates of strategies nicer (more likely to cooperate) or nastier (more likely to defect) than TFT were different. The nicer strategies often did not respond effectively to their opponents' defection—they were wiped out by the nastier strategies, which took advantage of the others' kindness. After the nicer ones were finished, however, the nastier ones began to suffer from low-scoring matches when they played against one another or TFT. Of course, some TFTs died out along with the nastier strategies, but more TFTs were produced by mutual cooperation when they played with copies of themselves. This was why TFT ultimately flourished.

Finally confident in the strength of TFT, Axelrod reflects, "What accounts for TIT FOR TAT's robust success is its combination of being nice, retaliatory, forgiving, and clear. Its niceness prevents it from getting into unnecessary trouble. Its retaliation discourages the other side from persisting whenever defection is tried. Its forgiveness helps restore mutual cooperation. And its clarity makes it intelligible to the other player, thereby eliciting long-term cooperation."[14]

In a nutshell, the karma of TFT's invincibility lies in being nice, retaliatory, and forgiving. But Axelrod still shied away from the seemingly inevitable conclusion that TFT was unbeatable. Exuberance for TFT aside, he noticed a hitch: when it played against a similar strategy—one that rarely but occasionally defected—if that other one defected on a single move, then both could become locked into an indefinite train of mutual defection. More simply put, a single incident of defection could trigger an unending streak of retaliation. Interestingly, this side effect may explain how spite evolved in human societies and, more importantly, whether we can overcome it. This takes us back to Hu's murder case in the previous chapter, where the fangs of spite sank deep.

Hu found justification for the extremity of his act in the fact that he hadn't initiated the aggression. This poses a general question: why is a retaliatory response (such as "an eye for an eye" and "a tooth for a tooth") of the same scale or stronger commonly viewed as legitimate? Remember, an essential component of TFT is never to defect first—for the sake of simplicity, let's call it the first-shot rule. If TFT is indeed the optimal strategy in human interactions, then retaliation, an essential component of TFT, is part of the evolved package of the overall strategy. That is to say, when our partner violates the first-shot rule, we have no choice but to punish the violator.

This first-shot rule is broadly heeded in most, if not all, societies to justify vengeful actions. Martial artists in ancient Chinese legends regard it as a sacred law; revenge becomes legitimate only after one has been offended, and it is demanded to redress the injury. Such legends have served as a benchmark for the society-wide moral standard of justice for time immemorial. Even in today's Chinese schoolyards, children stick to this age-old convention. When a scuffle occurs, each child often accuses the other of starting it, or, in other words, of violating the first-shot rule. Criticism or blame is more likely to adhere to those who initiate conflict rather than to those who respond. Brawls in the street, in a neighborhood, and between a couple are similarly resolved. Arbitrators commonly ask who has initiated the dispute before deciding on who is right and who is wrong. The rule is also used by the government to justify its actions in international conflicts. In recent memory, it was applied by the Chinese government to the military clash with the Soviet Union in 1969, the territorial skirmishes in Vietnam, Malaysia, and the Philippines in the South China Sea in the 1970s, and the brief war with Vietnam in 1979.

As in Eastern societies, the first-shot rule is followed widely in the West. In America, people often make the accusation "He did it first!" to justify their vengeful actions, in legal proceedings as well as in less formal social situations. The same claim is also heard among children for their retaliatory responses. Nations behave much like individuals in this regard. For example, on December 7, 1941, the surprise attack on Pearl Harbor by the Japanese Imperial Navy was a turning point in US history, prompting the United States to declare war on Japan and enter World War II after a protracted period of indecision. Sixty years later, in an astonishingly similar manner, the 9/11 attack by al-Qaeda instigated a nationwide mobilization to fight terrorism, and, as a result, the United States entered wars in both Afghanistan and Iraq.

Why is the first-shot rule broadly observed by people in Eastern, Western, and even isolated tribal societies? More to the point, what is the origin of this apparently universal penchant to obey the first-shot rule? From the perspective of game theory, TFT mandates that when your partner defects—that is, violates the first-shot rule—you must retaliate. If you don't, you take a net loss in relative payoff compared with your partner. Likewise, underreaction isn't an option, either. It will allow your partner to lose less to you than you to your partner. Again, you will suffer a net loss relative to your partner. So, once the first-shot rule is violated, we end up in spite, according to the dictates of TFT (see table 5). However, abiding by the rule has the distinct advantage that it can divert us from spite in the first place, especially when mutual destruction is not in the best interest of the parties involved. Thus, it should come as no surprise that the first-shot rule has been codified in human societies across the globe.

Retaliation—also a vital element of TFT—is a strategic necessity when the first-shot rule is violated. The instinctive vengeful response in humans, especially in young children, is evidence that revenge is literally in our blood. Research has demonstrated that our bodies are primed to seek recourse in revenge when we feel offended. The level of the stress hormone cortisol in the bloodstream jumps at the first sign of offense and stays high for a prolonged period of time if the conflict is unresolved.[15] The nucleus accumbens in the brain—an area related to Schadenfreude, the gratifying feeling of revenge—becomes activated when people who have offended us are treated unfairly. This activity becomes stronger as the yearning for vengeance intensifies.[16] Revenge is indeed sweet: it satisfies our desires for justice, thus lowering our level of stress.

However, the sweetness of revenge is transient, as retaliation begets retaliation in kind. After the first-shot rule is violated, peace is destroyed. Formerly cooperating partners turn against each other and go down together in a never-ending spiral of spite, often winding up in tragedy. This was illustrated by the Hamilton-Burr duel, one of the most dramatic climaxes of an interpersonal conflict in American history. Before the fatal event on July 11, 1804, the ratcheting up of the conflict had been proceeding for quite some time. The cycle of spite is highlighted by historian Joseph Ellis:

> In 1791 Burr defeated [the incumbent] Philip Schuyler, Hamilton's wealthy father-in-law, in the race for the United States Senate . . . . It was all downhill from there. Burr used his perch in the Senate to oppose Hamilton's fiscal program, then to decide a disputed (and probably rigged) gubernatorial election in New York against Hamilton's candidate. Hamilton, in turn, opposed Burr's candidacy for the vice presidency in 1792 and two years later blocked his nomination as American minister to France. The most dramatic clash came in 1800, when Burr ran alongside Jefferson in the presidential election . . . . The election was thrown into the House of Representatives because of the quirk in the electoral college . . . which gave Burr and Jefferson the same number of votes . . . . Hamilton lobbied his Federalist colleagues in the House to support Jefferson over Burr for the presidency, a decision that probably had a decisive effect on the eventual outcome. Finally, in 1804, in the campaign for governor of New York, which actually produced the remarks Burr cited in his challenge, Hamilton opposed Burr's candidacy for an office he was probably not going to win anyway.[17]

A crossover between their public and private lives served to aggravate the relationship between Hamilton and Burr. As such, political discourse metamorphosed into a showdown of personal honor, and political debate became enmeshed with bitter personal attacks. Worst of all, a duel was poorly conceived as a fair solution to end the fifteen-year-long history of animosity between the two.

Though conflicts driven by revenge and retaliation are rarely as dramatic as the vendetta between Hamilton and Burr, arguments between couples, friends, or coworkers can escalate in the same manner, often triggered by trivial issues. Such conflicts are so common that they are hard to avoid noticing in many aspects of our lives. Conflicts between parties, factions, bands, communities, states, and nations—driven by revenge and retaliation—can follow a similar path of progressive deterioration. Often they continue to escalate until a violent or tragic finale occurs. Hence Milton's observation, "Revenge, at first though sweet, / Bitter ere long back on itself recoils,"[18] is truer than its popular truncated formulation, "Revenge is sweet."

Revenge and retaliation can be contagious, and so they may be better viewed as having some qualities in common with a disease.[19] When our own interests are encroached on by our partners, we view it as betrayal—a violation of the first-shot rule. Often this perceived injustice will goad us into taking vengeful actions in an attempt to get even or, better yet, to reduce the relative advantage of our partners to below that of our own. Studies show that even the mere thought of someone who has hurt us can make our blood pressures rise and hearts beat faster.[20] These findings are further evidence that revenge and retaliation are indeed in our blood, and they elucidate why it is so hard to harness our vengeful rage when we feel double crossed. But if we follow our impulses, we are almost guaranteed a worse reaction. A vicious cycle of hostility begins.

Why do revenge and retaliation often proceed like self-fulfilling prophecies with rather predictable results? Usually, the cycle begins with a trivial event—a misunderstanding, a squabble, an accidental violation, or even a well-intended remark. Once it is taken as intentional or malicious, a vicious cycle of spite may ensue. Once the exchange of hostilities passes a tipping point, often no resolution except mutual destruction appears possible. Is there anything to pull us out of such a doomed cycle? Obviously, the ancient Greeks were already grappling with the same problem. One answer, offered by Aeschylus in his tragedy *The Oresteia*, is that "there is pain enough already. Let us not be bloody now."[21] Can such a forgiving strategy prevail? The answer is a cautionary yes: it can be done, but not easily.

Forgiveness is restraint from revenge and retaliation or an intentional underresponse to defection. As TFT demonstrates, if one of the two internecine parties takes the lead to forgive, the other party will likely follow suit. When this scenario occurs, the vicious cycle is broken, peace is restored, and the parties have a fresh start for cooperation, mutual tolerance, and peaceful coexistence. Forgiveness, therefore, may have evolved as an antidote to pull us out of the downward-spiraling maelstrom of revenge and retaliation.

As tools capable of breaking the vicious cycle of spite, acts of forgiveness have been observed in many social mammals, particularly in primates. Most species of monkeys and apes show some form of reconciliation after a fit of physical conflict. Lengthy grooming aside, baboons grunt, bonobos engage in sex, and some macaques expose their rumps to each other. Far from the apparent connotation of eroticism, these are behaviors intended for reconciliation, indicating that peace, not war, is welcome.[22] To serve the same purpose, humans have evolved a dazzling array of peace-making behaviors from handshaking, hugging, kissing, feasting, having make-up sex, to elaborate communal rituals.

Since forgiveness can break the spell of deeply entrenched spite, mend a wounded relationship, and restore cooperation, it's the reason why "turning the other cheek"—an apparently risky strategy—occurs in human interactions. Surveys in the United States two decades ago showed that forgiveness was the fourth most desired personal quality, behind honesty, ambition, and a sense of responsibility.[23] When asked whether they would forgive deliberate offenders, 45 percent of the participants said they would like to give it a try.[24] When offenses were severe, however, more and more people leaned toward revenge rather than forgiveness. A *Time* magazine survey in 1999 showed that only 20 percent of people would choose forgiveness if they or their family members were raped and only 15 percent would do so if their children were murdered.[25] For the same reason, a good number of Chinese, Koreans, and US veterans traumatized by the brutality of the Japanese military during World War II continue to hold feelings of ill will toward the Japanese. Forgiveness seems best suited for minor offenses.

Forgiveness, tolerance, and nonviolence are considered virtues in nearly all major ancient religions and philosophies: Confucianism, Buddhism, Hinduism, Judeo-Christianity, and Islam. That these are coping strategies for resolving the paradox of spite, for which they can be adaptive, is too obvious to miss. The nonviolence movements in the twentieth century exemplified by Gandhi, Martin Luther King Jr., Corazon Aquino, and Nelson Mandela arise from the ancient wisdom of forgiveness. Nonviolence, by restraining from or renouncing retaliation, is a sacrificial action that openly defies the response dictated by TFT. Gandhi, in particular, became a towering historical figure for his use of nonviolence to accomplish political goals. Awed by the power of nonviolence, he left the indelible line in his autobiography, "When I despair, I remember that all through history the way of truth and love has always won. There have been tyrants and murderers and for a time they seem invincible, but in the end, they always fall—think of it, always."

Gandhi was so confident in and committed to nonviolence that he challenged the ancient principle of retributive justice by stating that "an eye for an eye makes the whole world blind." Before Hitler's genocide of the Jews and others, Gandhi offered this advice to Jewish people, "If I were a Jew and were born in Germany? I would claim Germany as my home, and challenge him [the Nazi German man] to shoot me or cast me in the dungeon; I would refuse to be expelled or to submit to discriminating treatment? If one Jew or all the Jews were to accept the prescription here offered, he or they cannot be worse off than now. And suffering voluntarily undergone will bring them an inner strength and joy."[26] Even in 1940, immediately before the anticipated Nazi invasion of Britain, Gandhi was still relentlessly repeating his forgiving, nonviolent approach. "I would like you [the British people] to lay down the arms you have as being useless for saving you or humanity. You will invite Herr Hitler and Signor Mussolini to take what they want of the countries you call your possessions. . . . If these gentlemen choose to occupy your homes, you will vacate them. If they do not give you free passage out, you will allow yourselves, man, woman, and child, to be slaughtered, but you will refuse to owe allegiance to them.[27]

In retrospect, this advice is naïve, offensive, and absurd. If nonviolence is indeed an effective way of breaking the vicious cycle of spite, why do we find Gandhi's recommendations unacceptable in this case? To put it differently, why do revenge and retaliation, as a means of self-defense, appear to be the only sensible choice under these conditions?

Here again, it takes two to be fair. Forgiveness can be highly effective only if both parties have the intention of ending the current hostility in order to look for a new and better start. This is exactly what we often observe and what we practice ourselves when resolving conflicts between friends, couples, coworkers, and business partners. Such a process can be especially effective with the involvement of a mediator to help reestablish communication and dispel misunderstandings.

Although many religions such as Buddhism and Christianity promote the merits of mercy and forgiveness, unilateral, unconditional forgiveness is unlikely to flourish. It cost Alexander Hamilton his life when he violated the principle of an eye for an eye during his duel with Aaron Burr. Even American folk wisdom discourages it, "Fool me once, shame on you; fool me twice, shame on me." Indiscriminate pacifism, as adopted by some religious sects such as the Quakers, the Jehovah's Witnesses, Mennonites, and several other descendants of Anabaptism, proclaimed American theologian Reinhold Niebuhr, "[is] a parasite on the sins of the rest of us, who maintain government and relative social peace and relative social justice."[28]

Niebuhr's harsh criticism is predicated on the risk of unconditional clemency—if you forgive when your opponent does not, you will suffer a great deal. Think about the time before Hitler carried out his genocidal policy, his self-styled Final Solution. If the Jewish people had staged nonviolent actions such as strikes, sit-ins, fasting, or any other forms of civil disobedience advised and practiced by Gandhi, they would have only given Hitler another excuse to execute them en masse. In fact, the large majority of European Jews did choose submission instead of fighting back when the Nazis rounded them up. Once in the concentration camps, they had forfeited critical opportunities to muster resistance. Many surviving Jews have regretted this compliance ever since. By the same token, peaceful protesters can suffer greatly from ruthless repression, as demonstrated in Hungary in 1956, Czechoslovakia in 1968, and Beijing in 1989. The practicality of forgiveness is not boundless, nor are the fruits of nonviolent resistance guaranteed.

The risk of forgiving is well illustrated by the Hamilton-Burr clash. During the fifteen-year history of animus between the two, there were numerous exit points where both might have apologized, forgiven the other, and reconciled their relationship. Yet neither took the initiative. If it was hard for Burr because of his reserved personality,[29] the more outgoing Hamilton might have made the first move, but he didn't—unfortunately. Hamilton thought that publicly retracting his negative opinion of Burr would make him appear dishonest. As the conflict escalated, both parties found it increasingly difficult to compromise. Even when challenged to a duel, Hamilton was unwilling to back down. "If he did not answer Burr's challenge," writes Ellis, "he would be repudiating his well-known convictions, and in so doing, he would lose the respect of those political colleagues on whom his reputation depended. This would be tantamount to retiring from public life."[30] Hamilton had to stand up to Burr's bellicosity, for tough talk needed the backing of action. At this stage, the stakes were too high for either to back down; the Rubicon of spite had been crossed.

Hamilton was not utterly consumed by the desire for revenge. Even during the duel, he still held on to the hope of salvaging peace through forgiveness. "What is virtually certain, Hamilton fired first and purposely missed."[31] But in doing so he made a fatal mistake—he failed to communicate to his opponent that he did so intentionally, as a stance to observe the first-shot rule. As a result, Burr took Hamilton's deliberate miss as his own good luck and responded with the mortal shot that ended Hamilton's life. Burr won the duel, but not the approval of American public. (Worse, he was later engaged in shady business in the trans-Mississippi territory with British authorities, for which he became the Benedict Arnold of the time.[32]) There was no winner in this case, perfectly illustrating the vicious cycle of spite.

Luckily, Hamilton's mistake was not to be repeated in another famous duel in American history, the nuclear brinksmanship between the United States and the Soviet Union known as the Cuban Missile Crisis in October, 1962. At the knife edge of nuclear warfare between the two superpowers, John F. Kennedy stuck to the first-shot rule, despite his public bravado in appearing willing to risk an all-out war against the Soviet Union. Even when the Joint Chiefs of Staff all recommended a full invasion of Cuba, he opted instead for a naval blockade in the Caribbean to foil the infiltration of Soviet missiles. The blockade wasn't entirely successful; several Soviet ships slipped through the naval cordon, but it signaled unequivocally that the United States would not violate the first-shot rule. The Soviets, under the leadership of Nikita Khrushchev, got the message and quietly backed down. Mutual restraint allowed diplomatic negotiations to take center stage, and the crisis was resolved peacefully.[33]

If the use of forgiveness in order to achieve peace, harmony, and cooperation has a limited effect and is not effective in every situation, what other initiatives can break the vicious cycle of spite? The answer is interventions from external agents, such as arbitrators, community leaders, law enforcement officers, and divine beings (in Christian societies, the teaching that revenge is God's prerogative has the implicit adaptive function of taming human impulse for revenge[34]). These are some of the major external sources whose sole purpose is to settle disputes and, more importantly, to prevent conflicting parties from again sliding into the vicious cycle in the future.

Hu's murder case couldn't be more different from the Hamilton-Burr tragedy, but it shares with that tragedy a common circumstance—a failure of the official channels responsible for enforcing the law. At the time Hamilton was killed, dueling was already outlawed in many states, including New York, where the duel took place. The law, however, was not strictly enforced, providing some room for the tradition to continue. Ellis discerned a patterned cultural backdrop behind the perpetuation of the *code duello*, writing, "Not that it would ever die out completely, drawing as it did on irrational urges whose potency defies civilized sanctions, always flourishing in border regions, criminal underworlds, and ghetto communities where the authority of the law lacks credibility."[35]

Generally speaking, weak law enforcement encourages wild justice. There is an abundance of examples from the Appalachian region, Kentucky in particular, from the late nineteenth century. Besides the well-known feud between the Hatfields and the McCoys, notorious vendettas of similar scale included those between the Turners and the Howards, the Frenches and the Eversols, the Martins and the Tollivers, and the Bakers and the Howards. Anthropologist Keith Otterbein discovered several peculiar similarities among these five feuds: in the heat of the conflicts, each involved about ten rounds of aggression, resulting in around thirteen deaths.[36]

"When one family fights with another, it's a feud." Writer Malcolm Gladwell laments, "When lots of families fight with one another in identical little towns up and down the same mountain rage, it's a *pattern* [emphasis in original]."[37] Some elements emerge from this pattern. First, the blood feuds were all triggered by a conflict of interest, emerging from issues as trivial as a debt dispute in the Hatfield-McCoy case or, with the Turner-Howard feud, an accusation of cheating in a poker game. Second, they all culminated in tragedies for those involved. The commonality was weak law enforcement, often tainted by favoritism in local governments. When law has little power to curb the vengeful impulse, jungle law—in the name of honor—triumphs. As a result, the feuds were sustained for decades. In the case of the Baker-Howard vendetta, it lasted well over a century, from the early 1800s to the 1930s.

The same pattern occurs in many American inner-city neighborhoods where law enforcement is weak. The code of the street prevails, and the crime rate, especially retaliatory homicide, is high.[38] In the absence of law, revenge and retaliation are frequently the only practical means for people to defend their own safety, interest, and reputation. Similarly, in close-knit societies, honor is often a matter of survival, without which it's hard to prevent others from taking advantage. People may defend their honor at all costs, including life.[39] Thus, a culture of honor is in effect a culture of Darwinian struggle spawned by lawlessness. Here, TFT prevails, and we are returned to medieval times, when revenge was a way of life.

With honor taking center stage, spectators often intensify the response to provocations. In fact, adding a single bystander can double the chance of escalation from verbal argument to physical fighting between two men in the street.[40] This bystander (or spectator) effect is familiar to anyone who has attended sports competitions such as football or soccer games. It is behind such phenomena as home-field advantage and the employment of professional cheerleaders.

In Hu's case, since everyone in the village was aware of the unprovoked attack on him, the bystander effect was maxed out. He would have lost all honor had he completely backed down from the conflict. It is understandable that Hu rejected his enemy's proposal for a private settlement—a stance of asking for forgiveness. How could he save face in front of the villagers if he took the money? Even if Hu wanted a compromise, what if the village leaders went back on their word? With their political clout and connections with higher authorities, they could come after him again.[41] One way or another, Hu had to settle the score once and for all to ensure his social survival in the community. The best outcome would have been that the government put all of his enemies behind bars—this was exactly what Hu first attempted. But when the authorities repeatedly failed him, Hu's options for legal retaliation had run out; the ratchet of spite had clicked to the last cog. If the government was unable or unwilling to act, as it seemed, he had to finish the job himself. The government's failure to intervene represented a grave negligence of duty, which in this case pushed Hu over the edge.

With little faith in the Chinese legal system, Hu was convinced that the bloodshed would continue. Compelled by the ancient Chinese logic of "pulling out the weeds by eliminating the roots," he believed he had to take out the scions of his enemies to protect his own child from future reprisal. Such calculated and forward-looking action made the massacre all the more macabre. The case became depressively tragic not because Hu was a frenzied maniac but because his reaction was meticulously reasoned and planned, and because it was executed with such seemingly cold blood.

Hu's tragedy also tells us that forgiveness, aside from its efficacy, is often a weak strategy. Without being backed up by the threat of stronger measures, forgiveness alone can be inadequate to safeguard against the returning cycle of spite. In other words, though forgiveness has the power, at least in some situations, to short-circuit spite, it is often too feeble to do the job. Though in the end forgiveness has its place, the vengeful aspect of TFT, for better or worse, may have the upper hand.

Violence, akin to Hu's brutal revenge, has rocked many nations in schools, on college campuses, in workplaces, and at shopping malls. Unfortunately, America takes the lead in gun violence in these public venues. The US Bureau of Labor tallied 421 workplace shootings in 2008 alone and an average of 564 work-related homicides each year from 2004 to 2008.[42] Although the details vary from case to case, the underlying causes often appear strikingly similar. Like Hu, the perpetrators tend to lack any previous troubles with the law. They often view themselves as victims of some injustice, such as delayed promotion or other forms of workplace mistreatment or social slight. Much as in Hu's village, the situations have often lingered for quite some time without being addressed. This enables the "get-even" mentality to fester and eventually prevail. Also like Hu, perpetrators of random violence tend to see no other solution but to serve rough justice themselves. Almost invariably, they're aware that their actions comprise unredeemable crimes and, in the end, commit suicide.

This profile of wild, vengeful justice fits well with the 1999 carnage in Columbine High at Littleton, Colorado, where twelve students and one staff member were killed, and the 2007 Virginia Tech massacre, where twenty-seven students and five professors were slaughtered. In the former case, the two killers, Eric Harris and Dylan Klebold, were initially described as anything but brutal. "The boys," writes Dave Cullen in his book *Columbine* "were both gifted analytically, math whizzes and technology hounds." These are typical American youngsters we see in our schools, in our neighborhoods, or even in our own households today. In addition, their academic records were solid. Eric "was a gifted student taking a pass on college" and "Dylan had a bright future . . . heading to college."[43] Harris, the leader of the two, was depressed for a while and was probably psychopathic as well. But what finally dragged them into the killing spree appeared to be a combination of factors in their personalities, behavioral problems, preoccupation with guns, bombs, violent movies, and video games, and the perceived injustice of a society rife with, in their perception, "stupid people." Apparently, the tipping point came when they were arrested for the petty crime of robbing an unattended van, their first brush with the law, a year before their butchery. Both were totally consumed by intense feelings of vengeance against society and the people around them. During the rampage, they screamed out the motivation behind their lethal action, "It's a revenge!"[44]

The more deadly Virginia Tech massacre took nearly the same path. The twenty-three-year-old killer, Seung-Hui Cho, openly declared his rampage an act of vengeance. Shy due to an anxiety disorder and awkward in speech, he was reportedly a victim of bullying in middle school. While majoring in English at Virginia Tech, he might have shown some signs of schizophrenia. But what truly pushed him over the edge was his negative college experience. He allegedly had troubles with romantic relationships and was accused of stalking two female students, for which he was ordered to seek psychiatric treatment, though he never received it. These psychological problems were apparently exacerbated by his perception of the economic inequality in American society, which was distinctly more polarized than Cho's native South Korea. In videos, he compared himself to Jesus Christ, with the implicit mission of serving social justice by vanquishing the rich. He admired the Columbine High killers, calling them "martyrs," and, in his recorded video manifesto sent to MSNBC, he aimed his hate-filled rants against the wealthy:

> You had everything you wanted. Your Mercedes wasn't enough, you brats. Your golden necklaces weren't enough, you snobs. Your trust fund wasn't enough. Your vodka and cognac weren't enough. All your debaucheries weren't enough. Those weren't enough to fulfill your hedonistic needs. You had everything.[45]

These words are unambiguous evidence of the injustice perceived by Cho. Unfortunately, while we often stereotype perpetrators of school and workplace violence as stressed, distraught, psychotic, neurotic, unstable, psychopathic, or simply demonic, we tend to overlook one fundamental motivator of their extreme actions: the perception that fairness and justice have been unserved for a prolonged period of time.

Revenge and retaliation, while often devastating at a personal level, can come to dominate our collective national mentality and determine our dealings with other nations. Sociologist E. L. Moerk finds that in every major international conflict in US history, from the Spanish-American War in 1898 to the invasions of Iraq and Afghanistan in the wake of 9/11, the run-up to the war is propelled by a sentiment that our nation fell victim to unprovoked attacks. We, under the influence of TFT, must respond in kind. Accordingly,

Michael McCullough poses several serious rhetorical questions regarding these conflicts:

> Could there have been a Spanish-American War without the sinking of the *Maine*? Could there have been a World War I without the sinking of the *Lusitania* and the supply ships the United States was using to send weapons to Great Britain, France, and Russia? Would the United States have had the will to enter World War II without Pearl Harbor? Would the United States have allowed itself to become mired in Vietnam without the Gulf of Tonkin attack?[46]

Behind these historic events is the same ancient motivation that was illustrated by Hu's tragic ordeal. The desire for revenge reflects an evolutionary past that has a continuing hold on our individual psyches and is a powerful force in society at large. The same mentality leads to revenge-motivated violence in schools, on campuses, in workplaces, at shopping malls, and in many other public venues. In this sense, we are still suffering the painful mismatch between our biological nature and our cultural institutions.

If our behavior keeps regressing to the essential elements of TFT, does it mean TFT is indeed unbeatable? After the famous computer tournament, Axelrod had a sense that TFT was not as robust as it appeared to be.[47] This was the reason he avoided declaring TFT the optimal strategy in the Iterated Prisoner's Dilemma Game. His unease with TFT was confirmed a decade later by two evolutionary mathematicians, Martin Nowak and Karl Sigmund. They showed that in a noisy world like ours, where communication is imperfect (that is, misunderstanding can and often does occur), an improved version of TFT with forgiveness about a third of the time, instead of invariant retaliation, can do better than TFT alone. Fittingly, they named such a strategy, Generous TFT or GTFT for short.[48] Patrick Grim, at the State University of New York, was even more optimistic. After adding in some spatial complexity, he found a strategy that can beat GTFT simply by doubling the rate of forgiveness in GTFT.[49] In 1995, Axelrod, working with Wu Jianzhong, a scholar from the Chinese Academy of Sciences, also discovered a new strategy that can outsmart TFT. All these improved versions of TFT share a feature—they are more forgiving than the original and therefore can pull the vicious cycle of spite out of its downward maelstrom before it sinks too deep. In Axelrod's own words, TFT "and other nice rules require for their effectiveness that the shadow of the future [for cooperation] be sufficiently great . . ." and TFT "is not forgiving enough."[50] The verdict is clear: TFT is too vengeful to be optimal in the real world.

We may finally feel relieved to learn that TFT has been dethroned, and strategies with more ingredients of forgiveness will eventually win out. This is encouraging, as forgiveness is an evolutionary gospel that can often prevent humans from becoming tangled in cycles of spite with our spouses, friends, partners, colleagues, and associates. Forgiveness, though not invincible, is the first—and also the best—line of defense against spite.

Optimism aside, we should remain aware that though the championship aura of TFT may have faded, the ingredients of TFT—short memory, revenge, and forgiveness—remain the same. What has changed in the improved versions of TFT is nothing more than the relative proportions of these ingredients. We can expect that as we adapt to our shifting cultural environments, new versions of TFT will continue to emerge and evolve. This is by no means to say that we are inevitably heading toward a more benevolent future. On the contrary, crime, political violence, and revolutionary calamities can take hold when our social and cultural institutions are out of sync with our desire for equality and fairness, a grave concern we will explore in the next chapter.

Chapter 9

FIRE BEHIND REVOLUTIONS

The Manifesto of the Communist Party, written by Karl Marx and Frederick Engels, was published in 1848. It marked the birth of Communism and the appearance of a major new player on the world's political stage. Calling for an end to private property and introducing the now famous credo "from each according to his ability and to each according to his needs," Communism aspired to turn Thomas More's utopian dream into reality. It was a bold new initiative demanding social, economic, political, and legal justice in a world dominated by "aristocratic elites."[1] Claiming that "the history of all hitherto existing society is the history of class struggles,"[2] the *Manifesto* calls for working people—the proletarian class—to rise up and lead the struggle against their rich capitalist exploiters and oppressors. It openly advocates the use of violence to seize power in order to build an ideal society under, as Lenin later advocated, "the revolutionary dictatorship of the proletariat." The message, at first glance, is alluring, compelling, and commanding—at least if you're a member of the oppressed proletariat.

The first to answer the call were the Bolsheviks, the Soviet Communists. On October 25, 1917, they staged an armed uprising in St. Petersburg, then the capital of Russia. A handful of sailors on the Baltic Fleet cruiser *Aurora* harbored near the city and a small band of leftist Red Guards launched an attack at government offices. Encountering little resistance, the rebels moved on to take the headquarters of the ruling regime in the Winter Palace, guarded by only a small number of Cossacks. They again prevailed with ease, toppling the Provincial Government headed by Alexander Kerensky.

What the Bolsheviks overthrew wasn't a Czarist government but the interim administration installed after the February Revolution eight month earlier, when Czar Nicolas II was forced to abdicate. The famous October Revolution was, in realty, a power grab by the Bolsheviks. Historian David Priestland characterizes it as "a Bolshevik insurrection amidst a radical populist revolution."[3]

In the early morning of July 16, 1918, the Bolsheviks secretly rounded up the Czar's family and shot all seven members—the Czar, his wife, and their five children—together with their family physician and servants. This marked the end of the powerful Romanov lineage that had ruled Russia for three centuries, since its founding in 1613. Under the leadership of Lenin, the new Soviet government nationalized banks, repudiated foreign debts, and confiscated factories and church properties. The blueprint for a Communist state was realized in Soviet Russia seventy years after the publication of the *Manifesto*. A few years later, in 1922, the vast Communist empire we know as the Soviet Union was formed, incorporating the Ukrainian, Byelorussian, and Transcaucasian Soviet Republics.

Enthralled by the success of the October Revolution, Chinese radicals tried their luck. They organized the Chinese Communist Party in 1921 and launched their own movement, hoping for a quick victory as well. But the bulwark of the wealthy Chinese ruling elite proved to be stronger than its Russian counterpart. In 1927, the Nationalist government ruthlessly repressed the Communist movement, butchering thousands of activists in Shanghai, Canton, Amoy (Xiamen), Changsha, Wuhan, Guilin, Ningbo, and other cities. The surviving revolutionaries fled the metropolises, took shelter in remote rural areas, and continued to resist by means of guerrilla warfare. Besieged by the relentless mopping-up operations conducted by the Nationalists, the Communists teetered on the edge of utter defeat. In the end, they were forced to trudge three thousand miles on a journey that has come to be known as the Long March, relocating their base from East China to the northwestern province of Shaanxi. With the Nationalists constantly nipping at their heels, 90 percent of the Communist troops were lost during the year-long struggle between 1934 and 1935. Luckily, the Japanese invasion of China in 1938 diverted the Nationalist forces from civil conflicts. The Communists survived, recuperated, expanded their numbers and strength, and then eventually came back with a vengeance.

By the time the Japanese surrendered in 1945, following the atomic bombing of Hiroshima and Nagasaki by the United States, the Communists had already consolidated and expanded their rural bases in northwestern China. They had recruited and trained a formidable armed force—the People's Liberation Army. Although massive numbers of Nationalist troops were hastily deployed to fill the power vacuum left by the Japanese in the Northeast—equipped and supported by the United States—they were still overpowered by the Communists. The People's Liberation Army, albeit poorly fed and badly equipped, crushed millions of Nationalist troops and drove them from the mainland to the island of Taiwan in 1949.

The victory of Communism in the two nations—one with the largest landmass and the other with the largest population—sent a shiver through the West. The alarm call became even more disturbing when all of Eastern Europe turned "red" in the wake of World War II. In Western Europe, there was a vibrant and growing Communist movement as well. With the lower strata of society as its base in free elections, the Communist Party easily picked up a fifth to a quarter of votes in such countries as France, Italy, and Finland after World War II.[4] In India, votes from low-caste people put the Communists in power in the state of Kerala in 1957.[5] Cuba joined the red brethren in 1959. Many Western governments panicked: Could Communism be stopped? If so, how?

The United States led the response with military operations. "One of the main features of Eisenhower's Cold War strategy in the Third World," writes Priestland, "was the use of the CIA to stage coups d'état against nationalists deemed to be too close to Communism."[6] Direct military invasions aside, the United States was extensively involved in covert operations against Communists or left-leaning governments in dozens of Third World nations up through the 1980s. Fidel Castro claims there were six hundred attempts to assassinate him, "from exploding cigars to fungus-infected diving suits—and even to damage the supposed source of his charisma, his beard."[7] Despite Castro's unabashed exaggeration, the CIA's efforts were real and persistent. But, as we will see in the next chapter, such tactics often do more harm than good in the long run.[8]

Despite the West's dogged efforts and billions of dollars poured into fighting Communism, Communist regimes continued to thrive in the Soviet

Union, Eastern Europe, China, North Korea, Vietnam, Laos, Cambodia, and many countries in Africa and Latin America. In order to untangle the myths surrounding the strengths and allure of Communism, it is of little help to portray Communism as an unalloyed evil. Nor is it useful to credit the spread of Communism solely to the economic aid, ideological campaigns, or direct interventions from the Soviet Union—for "communist systems which have survived over the long term have never relied on either coercion or charismatic leadership alone."[9] On the contrary, as historian Archie Brown continues, "In a majority of cases they have had a base of mass support." The logic, at first glance, appears muddled when we observe that Communist states were typically plagued by poverty. When considered side by side with the enviable opulence of capitalist societies, how could they gather mass support, since their own economies were feeble and their living standards pathetic by comparison? If the pursuit of a comfortable material life is a universal human motive, how could wealth creation—the economic gospel of capitalism—be unpopular?

One of the first leads in cracking this vexing puzzle came in 1970, when sociologist Ted Robert Gurr published his book *Why Men Rebel*.[10] After examining a large number of revolutions and civil rebellions, including the American War of Independence in 1776, the French Revolution of 1789, the Russian revolutions in the early twentieth century, and the Spanish Revolution of 1931, Gurr discovered that a common denominator behind them was what he called relative deprivation (RD), measured with a simple formula:

$$RD = \frac{V_e - V_c}{V_e}$$

where v_e is what one expects (and most likely, the highest realistic goal) and v_c is what one currently has.[11] The two value terms, v_e and v_c, can be any of the desired conditions: economic well-being, social status, political rights, educational opportunities.

For illustration, let's put the French Revolution of 1789 under the lens of Gurr's RD analysis. Before the revolution, France was in an "incontestable expansion of agricultural, industrial, and commercial profits."[12] Fueling the economy were wage earners, many of whom were recruited from the landless peasantry. Yet factory workers, despite their better incomes than poor peasants *in an absolute sense*, lived on bottom-of-the-barrel wages. Compared with their peers in the countryside, they sat lower and deeper in the urban socioeconomic hierarchy. In Gurr's formulation, the gap between V_e and V_c widened, and, for the wage earners, the relative deprivation, RD, increased. Meanwhile, advances for those in the middle class—lawyers, financiers, merchants, professors, physicians, government officials—were arrested by the inherited privileges of nobles and clergymen.[13] The hazy popular desire for equality, fairness, and justice was sharpened by the writings of Enlightenment thinkers such as Voltaire and Rousseau. When the legitimacy of the law, the monarchy, the aristocracy, and the Catholic Church in the ancient regime was called into question, the tempest of revolution loomed on the horizon. Again, it was the gap between what people had and what they expected that gave impetus to the French Revolution. Hence, Gurr's idea of relative deprivation is none other than our general concept of relative payoff that we've encountered in other guises throughout this book—the gauge that measures the strength and motivates the expression of our fairness instinct.

This explains how the French Revolution could occur even in an expanding economy. "As a group experiences an improvement in its conditions of life, it will also experience a rise in its level of desires," diagnosed Alexis de Tocqueville. "The latter will rise more rapidly than the former, leading to dissatisfaction and rebellion."[14] As the economy grew in late-eighteenth-century France, the gap in social and economic status between the haves and have-nots widened. The feeling of depravation deepened, and the desire for equality and justice intensified among the poor and disadvantaged. Two measures might have been taken to soothe this searing demand for fairness. One, advocated by the Jacobin journalist François-Noël Babeuf, was to "eradicate once and for all the desire of a man to become richer, or wiser, or more powerful than others,"[15] an idea that would become a Communist ambition six decades later. The other was to remove obstacles—the inheritance of social status, the concentration of wealth, the privileges of the noble and the clergy—that hindered the middle and lower classes from social and economic advancement. Unfortunately, the ancient regime took neither of these paths to ameliorate the unrest. Complicated by an untimely financial crisis in 1789, France was caught in a perfect storm of violent revolution. As a result,

> a social order founded on legally entrenched and inherited hierarchy collapsed. The estate system was abolished, and with it the notion that men were born into particular and tiered stations of society ordained by God. No longer were the first two estates—the clergy and the aristocracy—to be privileged over the rest of society—the "third estate." All men were declared to be legally equal, "citizens" of a single, coherent "nation" rather than members of separate estates, corporations and guilds. In part, these demands for legal equality arose from third-estate anger at the superciliousness of the aristocracy; ordinary people also resented having to pay taxes from which their "superiors" were exempt.[16]

This seemingly counterintuitive link between economic expansion and popular revolt is by no means unique to the French Revolution of 1789; there are other instances of this phenomenon in France, Britain, Russia, and America, when societies were enjoying vibrant economic growth,[17] thanks mostly to capitalism. There is, of course, nothing wrong with economic growth. The culprit lies in the widening inequality that often comes with it.[18] Despite being a common byproduct of economic expansion, inequality acts as a divisive force that can dampen the spirit of social cooperation, destroy our sense of community, alienate us from shared values and responsibilities, and, when extreme, fuel hostility between the haves and have-nots. Consequently, inequality can instigate violence, destabilize societies, and pave the way for the collapse of states.[19]

The theory that economic inequality breeds political violence was known to Greek philosophers, Aristotle in particular. After the French Revolution, it appeared to be beyond debate. "Almost all of the revolutions which have changed the aspect of nations have been made to consolidate or to destroy social inequality," wrote Tocqueville in 1835 with an assuring certainty. "Remove the secondary causes which have produced the great convulsions of the world, and you will almost always find the principle of inequality at the bottom."[20]

Tocqueville's intuition has survived stringent statistical analyses and experimentation with real people. Numerous studies have demonstrated that people in the lowest economic strata of a society are most likely to revolt against existing sociopolitical systems.[21] High crime rates in many areas of American inner cities are sober statistical windows, through which we can see into the dark side of our fairness mentality. In societies with little social mobility, where the underprivileged are unlikely to advance, the risk of political violence rises with the level of inequality.[22] In a study using cross-national data, political scientist Edward Muller reveals a disturbing pattern: the more the top 20 percent of households earn, the more people die in political violence.[23] Marie Besançon of Harvard University shows that when economic inequality, as measured by the Gini index, increases from .28 to .45, the probability of revolution can rise five-fold.[24]

Norms of fairness are broadly practiced by people all over the world.[25] Blatant violation of these norms often instigates violence and revolts in addition to major revolutions. In America, a study in 1999 showed that income inequality was closely linked to violent crimes such as homicide when measured by the Robin Hood Index (similar to the Gini Index) across the fifty states.[26] In Latin America, known for high levels of economic, social, and political inequality, rates of violent crime such as robbery and homicide not only stayed high but held infamous world records in Colombia, El Salvador, Brazil, and Venezuela at least up until the 1990s.[27]

In dynastic China, "Even up the rich and the poor!" was a rallying slogan used in virtually every major peasant uprising from the first dynasty, the Qin, in the third century BCE, to the last dynasty, the Qing, in the nineteenth century. This happened regardless of vast differences in social, economic, and historical circumstances. The fact that this simple rallying cry could attract such broad support from the poor across the ages speaks to the power of fairness when resource distribution is extremely skewed. To appease public unrest and resentment, rebel leaders (and occasionally newly crowned emperors) forced top-down reforms by confiscating lands from the wealthy and redistributing them to the poor.

For the same reasons, major revolutions across the world are commonly incited by the desire for fairness and justice. The Revolution of 1905 in Russia occurred after a long, steady period of capitalist economic expansion. In a manner similar to the French Revolution of 1789, factory workers, despite the fact that their incomes were higher than serf-like rural farmworkers, became increasingly aware of their relatively low socioeconomic status in the city. This instigated labor strikes that at times could involve as many as 120,000 participants in St. Petersburg during the reign of Czar Nicholas II.[28] Even the more recent Iranian Islamic Revolution of 1978–79, generally considered a theocratic revolution, was at least partially accounted for by economic inequality.[29] In fact, the revolution occurred under the reign of Shah Mohammed Reza Pahlavi, when Iran's economy, buoyed by oil revenues, was among the most vibrant in the Middle East. With the economy completely ruined in its wake, the revolution served as an ironic example of society-wide spite between rich and poor.

The United States, too, has sailed its course through history guided by the compass of equality. The American War of Independence in 1776 was triggered by longstanding resentment of taxes (such as the Sugar Act of 1764, the Stamp Act of 1765, and Townshend Acts of 1767) levied on American settlers. The British government grossly miscalculated how indignant Americans were toward "taxation without representation." Historian David McCullough writes: "The Americans of 1776 enjoyed a higher standard of living than any people in the world. . . . How people with so much, living on their own land, would ever choose to rebel against the ruler God had put over them and thereby bring down such devastation upon themselves was for the [British] invaders incomprehensible."[30] But rise up the Americans did. Jefferson's political essays, *Summary View* and *Causes and Necessity*, were replete with strong sentiments about unfairness and injustice, condemning the authority of the British parliament and King George III over the American colonies. The most memorable and powerful phrases in the Declaration of Independence are refined from these ideas. "The demand for equality," points out political scientist Mark Lichbach, "has lain at the epicenter of the major upheavals that have erupted on the American political scene: the Revolution, the Jacksonian era, the Civil War and Reconstruction, the Populist-Progressive period, the New Deal and the tumultuous 1960's and 1970's. The general association of inequality with conflict thus appears inevitable and immutable."[31]

In case after case, it is difficult to avoid seeing a pattern: revolutions are most commonly triggered by the perception of inequality and injustice among the disadvantaged. As such, "equality," "fairness," and "justice" are cloaked in an aura of sacred power, for which they become popular rallying calls against the established order. The American Declaration of Independence exemplifies this in its popular phrase: "We hold these truths to be self-evident, that all men are created equal." During the French Revolution of 1789, "liberty, equality, fraternity, or death" summarized the irresistible zeitgeist. Not surprisingly, "equality," "justice," and "democracy" are among the most common words in the constitutions of many nations today. Even in countries with little freedom, democracy, or respect for human rights, such words are employed for decorative and rather cynical purposes. For instance, North Korea, officially known as the Democratic Republic of Korea, is a country where its founding leader, Kim Il-sung, was publicly worshipped as "superior to Christ in love, superior to Buddha in benevolence, superior to Confucius in virtue and superior to Mohammed in justice."[32]

The idea of Communism arises from the age-old custom of collective ownership and equal sharing of resources, a co-op system that, as Priestland shows us, has deep historical roots:

> The inhabitants of Plato's ideal "Republic" held property in common, and the early Church provided a model for fraternity and the sharing of wealth. This Christian tradition, combined with traditional peasant communities' cultivation of "common land," was the foundation for the Communist experiments and utopias of the early modern period . . . . But all of these projects were founded on the desire to return to an agrarian "golden age" of economic equality, whereas future Communists also claimed they were creating modern states based on principles of political equality.[33]

As might be expected, Communism, with its promise of social and economic salvation, was passionately embraced as a new messiah, especially by the poor—at least in its initial stage. "Communists did best in underdeveloped agrarian economies where industrialization was late and patchy, and the working class was poorly organized," writes Priestland. "In these countries, peasants tended to be angry at the remnants of an old agrarian order."[34]

To satisfy the popular desire for equality, Communist states, once established, imposed radical changes. Politically, they provided social mobility to those who were shut out of political power under the old order. Economically, they demonized private property as the source of all capitalist evils and staged intensive campaigns to minimize inequalities among people via nationalization of financial institutions, ending private property, building public housing projects, compressing income disparity, and collectivizing industrial and rural production. Socially, they provided mass education, universal healthcare, and job security with full employment. In addition, they took measures to homogenize lifestyles down to minute details such as instituting dress codes and codifying eating habits. In peasant societies with high levels of inequality, Communist parties catered to the material and emotional needs of the poor through land reform and persecution—to the degree of mass execution—of the wealthy.[35] "Communist parties, in the countries in which they came to power," observes Brown, "have been able to tap into patriotic and anti-imperialist sentiments as well as [in]to the desire of the poor to reduce inequality and take revenge on those perceived to be their class oppressors."[36] "Their emphasis on welfare, education and social mobility," adds Priestland, "was often in sharp contrast to the priorities of the rulers who went before, and could be very popular."[37]

This scenario fits perfectly with the success of the October Revolution in 1917, which followed enormous Russian casualties during World War I, with one million soldiers killed and four million wounded. Promising peace and land reform (redistribution), the Bolsheviks took advantage of the crisis and appealed to the downtrodden masses with a three-word slogan: "freedom, bread, and peace."[38] In his most important work, *The State and Revolution*, Lenin made a grand appeal for equal status for all: "only communism makes the state absolutely unnecessary, for there is *nobody* to be suppressed."[39] The message struck a chord.

Even in the West, Communism drew a good deal of sympathy and a significant following. During the times immediately before and after World War II, droves of people in capitalist England, Ireland, France, Italy, Spain, and the United States joined the Communist Party with the genuine belief that Communism would bring justice, peace, prosperity, and the end of human suffering. Jewish people were particularly attracted to the idea due to their history of oppression and marginalization in European countries. Even today, some Israelis still live communally on collective farms—the kibbutzim—without private property. For quite some time, it seemed the Communist system justified a destiny foretold by Marx, "Communism is the riddle of history solved, and it knows itself to be this solution."[40]

Inequality, regardless of its power to inspire violent revolt, usually will not ignite a full-bloomed revolution. More often than not, an organized revolution requires enlightened and well-educated leadership if it is to take hold. In "backward" societies, "Marxism's desire to unite modernity and equality was . . . appealing to the patriotic students and educated elites."[41] In this regard, Communist leaders such as Lenin, Trotsky, Bukharin, Chen Duxiu, Mao Zedong, Ho Chi Minh, Fidel Castro, and Che Guevara shared a generic profile. They all paid keen attention to the lives of the poor. Mao, for example, wrote several reports in his youth about the gloomy conditions of rural peasants in his native Hunan province. Driven by the desire for a better society, these leaders came to the same erroneous conclusion: only Communism can restore justice by filling the canyon between the haves and the have-nots.

Indeed, the Communist movement drew its strength from the desire for equality, primarily on the part of the poor. In its initial stages, it succeeded in inspiring those who wanted social and economic justice to join the cause. For example, Che Guevara, the radical Communist guerrilla fighter, was euphoric about how Communism might unleash human energy for building a better world. "Che's ultimate communist utopia liberated the individual from the sense of alienation experienced by workers in a capitalist setting. Each person, freed from the oppressive restraints of capitalism, would express his or her human condition through music, literature, and the arts, producing an egalitarian society in which culture and the arts flourished."[42] After visits to the Soviet Union and China, he was ecstatic about the potential of Communist nations. "Their strength, their high rates of economic development, the dynamism they show, the development of all the people's potential, convince us that the future definitely belongs to all the countries who struggle, like them, for peace in the world and for justice, distributed among all human beings."[43]

The Communist economic system, however, failed to deliver the promised results. In Communist states, the abolition of private property and the emphasis on absolute equality suppressed the evolved human desire for outcompeting the rest of the pack. With this motivation suppressed, mediocrity prevailed, and the vitality of the economy withered. Worse, planned economies often decouple supply and demand, leading to severe shortages in consumer products—such as toys, fabrics, utensils, clothes, and personal-hygiene products—despite the overproduction of certain industrial goods such as steel and heavy machinery. As a result, numerous economic niches go unfilled and services are typically terrible—a scenario familiar to anyone who lived in China or the Soviet Union before the 1990s. With an economy in extreme distress and poverty widespread, how could the Communist distribution principle, "from each according to his ability and to each according to his needs," be satisfied? To put it more directly, how can you meet demand when there is a paucity of supply?

More troubles emerged from the political system. The monopoly of power in Communist states, despite the Communist Party's goal of leveling economic status, inevitably led to inequality in the sociopolitical sphere. As privileges for party officials became widespread, economic inequality grew more prominent and, in some cases, exceeded that of the United States. As a result, the proletarian dictatorship, after eliminating the old aristocratic plutocracy, ushered in new forms of hierarchical bureaucracy that had their own deleterious socioeconomic effects. "Many Communists and even ordinary citizens . . . no longer expected it [Communism] to forge radically egalitarian social relations, or to create a dynamic new economy to compete with capitalism: both radical equality and economic dynamism were simply too difficult to reconcile with party dictatorship and the command economy."[44] In the end, autocracies or even dictatorships developed. Monopolies of power eventually led to the demise of the central promise of Communism: equality and justice for all.

Rigid political and economic hierarchies emerged in most Communist nations. Inequality became increasingly prominent from the 1920s in the

Soviet Union and the 1950s in China. In the ensuing decades, rations, wages, access to limited consumer goods and services, housing, and many other basic needs took on the status of privileges and were doled out according to the ranks of party officials. Status and income gaps between ordinary workers and managerial officials also widened considerably.[45] In China, for instance, government officials were classified into twenty-four ranks (in cities) or twenty-six ranks (in the countryside), with a salary differential between highest and lowest that varied by a factor of nine.[46] More conspicuous were privileges involving housing, traveling, shopping, vacationing, and other categories, meticulously delineated and meted out to specific officials of different ranks.

Few Communist leaders expected the dire aftermath of these power monopolies, despite Lord Acton's famous warning, "Power corrupts and absolute power corrupts absolutely." Corruption appears to be a chronic cancer in human society, and Communist society was not immune to it. In an attempt to counter this disease, the Chinese Communist Party made fighting corruption a top priority years before it took control of China in 1949. One classical example was the execution of Huang Kegong, a promising young leader and Mao's loyal subordinate, who killed a female student after a failed courtship. Huang wrote a touchingly remorseful petition to Mao, hoping Mao would intervene and spare his life. Mao, however, rejected Huang's plea so as to make himself an example against power abuse. He was said to have been in tears when Huang was executed.

As cases of corruption kept popping up, the measures taken by the Chinese Communist Party became harsher. By 1943, embezzlement, if the sum was over 500 yuan—a lot only in a relative sense—was punishable by death. In the early 1950s, the new Communist government staged a series of sweeping campaigns, known as the "Three-Anti" and "Five-Anti" movements, to purge party officials and businessmen involved in bribery, bureaucracy, wasteful spending, tax evasion, cheating on government contracts, appropriating state property, or stealing economic secrets.[47] In the late 1970s, the Chinese leadership created a new branch, the Central Commission for Discipline Inspection, whose sole duty was to fight official corruption. These efforts, however, had little palpable effect in curbing the abuse of power, a vice that was too deeply ingrained for the Communist government to wipe out.

This illustrates a serious shortcoming in Lenin's theory, which contains few ideas regarding the structure of institutions that can be used to buttress accountability and individual freedom.[48] Without checks and balances, how could a monopoly of power on the part of a single party (or worse yet, a party leader) be avoided? Without media censorship, party control at every government-run work unit, jamming of Western radio stations, and surveillance over expressions of personal opinions (including mail communication) by the secret police, how long could a Communist state survive?

Central among the numerous problems that arose in Communist systems was this question: why did Communist nations tend to be fraught with political repression? Stalin, for instance, degenerated into a ruthless dictator. He committed monumental crimes—the liquidation of rich peasants, the starvation of millions, the Great Purge, the Gulag. From 1937 to 1938, 681,692 Soviets were executed and over 1.5 million were imprisoned, mostly for "political" reasons.[49] When famines are added to the list, the number of lives lost under Stalin's reign exceeded ten million.[50] A large number of the victims were Communist revolutionaries themselves. Likewise in China, where power struggles among top leaders resulted in catastrophes such as the Great Famine of 1959–1962 and the Cultural Revolution of 1966–1976. The inviolable right of human life has never been the moral baseline of Communist rule.

Blissful ignorance of the power of human nature, a euphoria shared by many Communist leaders, was the reason they failed to create effective institutions to harness corruption and power abuse. It crippled the Communist system and ultimately doomed the hope that the Communist utopia would serve as an ideal system of justice among people.

With at least the perception of inequality banished, people living in Communist states were generally content with the status quo, despite stagnant economies and low living standards, often bordering on abject poverty. But such stark reality was hidden beneath propaganda campaigns and information isolation—sturdy firewalls that prevented people from comparing their living conditions with those in the West. The information firewall began to come down when the open door policy was initiated in China and the Soviet Union in the late 1970s. This made the vast economic gap between the Communist bloc and the West painfully obvious; the ill effects of the

Communist Party's monopoly on power became evident, as did the privileges reserved for the bureaucratic ruling class. These factors contributed to the downfall of Communism in Eastern European countries and the Soviet Union in the late 1980s and early 1990s. In Romania, the transition was so violent that its long-time Communist dictator Nicolae Ceausescu and his wife were executed on Christmas Day in 1989 by the military rebels, an event reminiscent of the massacre of Czar Nicholas II's family seventy-one years earlier.

As a result, no Communist state exists in Europe today. Among the six Communist states in the world, some, such as China and Vietnam, have abandoned the Communist economic system. In China, the new laissez-faire market economy has produced a Gini index higher than almost all traditional capitalist nations, including the United States,[51] Great Britain, France, and Germany. In a characteristically practical manner, leaders of the Chinese Communist Party seriously considered dropping the word "Communist" from its official name a few years ago.[52] Even the North Korean leadership accuses China of deviating from the Communist cause. In Brown's assessment, the "transnational [Communist] movement has gone, and so has the aspiration to build a communist society. In spite of lip-service to the goal of communism, no ruling Communist Party any longer places emphasis even in theory on movement towards the stateless society, the culminating and 'inevitable' stage of human development, as envisaged by Marx."[53]

What brought Communists to power was an enticing blueprint for a Shangri-La society where everyone would be equal, but Communism failed because of a reality it was unable to manage: socioeconomic inequality that arose from a political hierarchy and the Communist Party's unfettered monopoly on power. Ironically, the meteoric fall of Communism was dictated by precisely the same social force that was responsible for its dramatic rise: inequality, which can serve as both maker and breaker of a socioeconomic system. Looking back, Communism was a beautiful mirage; it offered a vision of equality and justice but proved to be an illusion, made unreal by the biological underpinnings of our fairness instinct.

If there is any enduring symbol of public romance with Communism, it has to be Che Guevara's image. Believing the fight for Communism was a just cause, Che willingly sacrificed all he had—wealth, power, comfort, privileges, family, children, and his own life—for Communist ideals. He led a multinational guerrilla campaign against the United States—viewed as the epitome of capitalist evil by many at the time—in Cuba, the Congo, and Bolivia. In 1967 he was captured and executed in Bolivia at the age of thirty-nine. Ruthlessness aside, Che was nonetheless broadly admired for his sense of justice. As a person, he was so self-disciplined and incorruptible that the French philosopher Jean-Paul Sartre extolled him as "the most complete human being of our age."[54]

In Western nations, Che became a countercultural icon in the 1960s and 1970s. He attracted a large following in America and Europe, especially among young people who revolted against the old order of imperialism, economic inequality, and social injustice. The Kennedy administration, keenly aware that economic and political inequalities could mobilize popular support for Communism,[55] was preparing to introduce major social changes. But it was largely under President Lyndon Johnson that sweeping initiatives for social equality were passed into law. As socialist programs such as Medicare and Medicaid were implemented and civil rights legislation promising racial justice was passed, America's ambition to become a "Great Society" looked ever more real; the societal tension and anxiety caused by inequality began to ease.

But Che's image popped up again two decades later. And this time, it was in China, a nation he visited twice in the 1960s, but where he was little known before the 1980s. Since the 1990s, however, Che's name has gone viral, allowing his legend to rub shoulders with the most popular idols in China. Some journalists claim that Che's fame exceeds even that of Lei Feng, a legendary native role model known for his sacrifice to the Communist cause. Notice that the 1990s was also a time when the Chinese economy began to take off and the gap between rich and poor started to widen. Was Che's rise to popularity in China during the great Chinese economic expansion a coincidence?

In 2001, the theatrical play *Che Guevara* premiered in many Chinese cities. The gala in the upscale Lanxin Grand Theater in downtown Shanghai drew a large crowd, a scene rarely experienced in recent years, since movies,

TV series, and reality shows began to dominate the entertainment business in China. Awed by Che's fame but unaware of who he really was, many among the nouveau riche and the Communist Party leadership also showed up . . . only to find themselves in the wrong place at the wrong time. Before the play drew to a close, many of them had already skulked out. While the media labeled the play "most controversial" and the premier "a major social event," those who were fed up with corruption and sickened by wealth disparity hailed it as "the most satisfying and delightful" drama in two decades.[56] It stirred in them the same feeling of injustice that inspired Che to fight.

This premier was soon followed by a rising tide of violent crimes against the rich. Among the victims was Minhong Yu, president and CEO of New Oriental Education and Technology Group, a public company listed on the New York Stock Exchange with a market capitalization topping $5 billion as of September 2011. He was abducted on August 21, 1998, and robbed again at his home in Beijing a few years later. Mr. Yu was extremely lucky to have escaped unharmed considering that several millionaires have lost their lives in the violent trend of "killing the rich," which has emerged since the early 2000s. According to an article on Chinanews.com, 3,863 cases of kidnapping were tallied nationwide in 2004 alone. Regardless of meticulous security measures taken around their homes, the rich are still unable to seal themselves off from society at large. Their vulnerability may come from the least expected places. In 2010, for example, dozens of children were brutally murdered and many more were injured in five unrelated assaults in kindergartens, primarily in upscale residential areas. A deep hatred of the rich was palpable.

It's ironic that the meteoric rise of Che's fame in China occurred when the Chinese economy was soaring. The absolute living standard has markedly improved, even for the bottom stratum of Chinese society. The poverty rate declined from 85 percent in 1981 to just 15 percent in 2005.[57] However, this economic miracle aside, the widening gap between rich and poor has touched the nerves of the Chinese public. In less than two decades, the collective attitude of that segment of the population that is relatively worse off has turned from admiration to envy and, finally, to active resentment of the rich. Today, luxurious cars parked on city streets are popular targets of vandalism, and riots have become commonplace in the countryside. In 2005, 87,000 protests were tallied nationwide. This statistic rose to 180,000 in 2010 and probably doubled again by 2012, according to an article in *Time* magazine.[58] According to the same article, the central government spent $110 billion, more than its entire defense budget of $106 billion in 2012, on *weiwen*, measures aimed at maintaining the stability of society. Never since 1949 has the ruling elite of the Chinese Communist Party faced civic unrest of this magnitude. And, paradoxically, it comes at a time when China's economic expansion appears unstoppable.

Though less dramatic, the Robin Hood mentality has been making itself felt more strongly in recent years within several capitalist nations where wealth has become increasingly concentrated among a tiny proportion of the population. Growing inequalities are almost a surefire recipe for political instability during economic expansions. In her 2004 book *World on Fire*, Amy Chua of Yale University rang the alarm regarding this phenomenon in many regions of the world. She shows, in case after case, how political violence can inspire hatred and resentment on the part of the poor majority toward the rich, privileged minority. For the same reason, the Communists have enjoyed a resurrection in India in recent years, as economic inequality grows while poverty remains ubiquitous among the masses. Incited by social inequalities in areas such as opportunity for education, Nepal became a Communist nation in 2008 via a popular election, countering the waning tide of Communism in the rest of the world.

In Che's Latin America, where wealth has been traditionally concentrated in the landholding class and social mobility is often extremely low, persistent socioeconomic inequalities have increasingly put Communist or left-leaning politicians in power. The victories in presidential elections of Hugo Chavez in Venezuela in 1998, Evo Morales in Bolivia in 2005, Fernando Lugo in Paraguay in 2008, and Dilma Rousseff in Brazil in 2010 are all consequences of these polarizing inequalities. In Mexico, where Carlos Slim Helú took the crown as the richest man in the world from Bill Gates in 2007, droves of poor rural peasants have continued to risk their lives to trek across the desert for opportunities in the United States. The Zapatistas in Chiapas, meanwhile, have followed Che's footsteps in their fight on behalf of the indigenous people under the name of "Marxist humanism."[59]

Che, as a revolutionary icon, has endured the relentless bleaching of time in the West. His famous image, taken by Cuban photographer Alberto Korda, lives on through T-shirts and posters decades later. BBC News reported on May 26, 2001, that the Maryland Institute of Art called the photo "the most famous photograph in the world and a symbol of the 20th century."[60] Songs like "*Hasta siempre, Comandante*" ("Farewell, Commander"), which was popularized by the American folk singer Joan Baez in the late 1960s and 1970s, met their modern version three decades later.[61] In 1997 the Spanish pop singer Nathalie Cardone's song with the same title became a hit worldwide. The five-minute-long MTV video starts at the run-down schoolhouse where Che was executed. The sexy singer is in a loose, old military shirt, her hair uncombed. Holding a baby (Che's heir), an AK-47 (Che's means for justice) strapped to her back, she passes through shanty towns, city slums, and cane fields where the blank-faced poor are living and working. The theme is interspersed with flashbacks of Che's images, and the lyrics end with a growing army of poor people, throwing down their work, following the singer, and marching along jagged streets and dirt roads in long files. The message is clear and powerful. The puzzling issue is why the song became so popular when the fate of Communism seemed sealed in history.

The answer is that—despite the fact that Che, along with the cause he fought for, is politically defunct—our inner Robin Hood instinct lives on. When inequality of any type becomes extreme, it reasserts itself. "The history of communism should have taught us two things. The first lesson . . . is how destructive dogmatic utopian thinking can be," Priestland concludes. "The second lesson, rather more neglected today, is the danger of sharp inequalities and perceived injustice—for they can make that utopian politics very appealing."[62] Despite the failure of Communism as a political and economic system, popular sentiment for Communism has never completely disappeared. The nostalgia for this failed ideology can be quite strong among the disadvantaged in China, Eastern European nations, and the former Soviet republics.

Although class struggle, in the Marxist sense, is an overly simplistic formulation regarding the relationship between rich and poor in society, the injustice felt by those on the lower rungs can nonetheless foster revolts and instigate uprisings. Even in established democracies, unrest due to perceived inequality, unfairness, and injustice often occurs. Indeed, there have been many riots in the United States, despite electoral and legal institutions considered exemplars among Western societies. In recent memory, the 1992 Los Angeles riots following the acquittal of the white policemen who beat Rodney King, an African American man, is such a case.

The revival of socialist and even Communist sentiments in America is noticeable in the wake of the bursting housing bubble in 2008. Reckless business practices calculated for maximum short-term profit, with little concern for the welfare of communities or long-term financial health, have been widely reviled and condemned. Wall Street, not long ago an emblem of capitalist triumph, has again become a symbol of capitalist evil. Bankers and CEOs, together with the super-rich, are once more the target of broad public resentment. One can only expect that, without the generosity of a few billionaires such as Warren Buffett and Bill Gates, "the 1%" would be even more detested by "the 99%" in America. The claim that capitalism is morally good appears to ignore the complex links among our economic and political systems, social stability, and our innate instinct for fairness.

A prison population of 2.3 million, along with an incarceration rate nine to ten times higher than that of many European nations, is just one indicator that the United States has already suffered a great deal from economic inequality.[63] A simple denunciation of Communism as an unalloyed evil can blind us from seeing the enduring appeal of this flawed ideology and the inherent weaknesses of free-market capitalism. Contrary to the overhyped misunderstanding in the United States that redistribution equals socialism or Communism, research has repeatedly shown that many programs, such as basic education, primary healthcare, and food security for the needy, will not harm economic growth or private investment.[64] On the contrary, they may enhance growth and investment. More to the point, these programs have enormous social value—improving stability, efficiency, labor quality, economic competitiveness, public safety, social mobility, and a sense of community—that is hard to measure in monetary terms.[65] Perhaps it's time to go beyond the political mudslinging that pits Communism and capitalism against each other so as to ask some probing questions. How can socialist elements in the United States such as Social Security and Medicare remain popular despite their existing problems? Does democratic socialism provide a middle ground by combining the economic strengths of capitalism (market economy) and social strengths of Communism (the promise of equality and social mobility) backed by the political principle of democracy? Or can we find a new system, beyond the capitalism-Communism continuum, that better fits our fairness instinct? Although answers to these questions may not come easily, it seems certain that there are better solutions than those offered by the status quo.

Che, like the medieval Robin Hood, has become more than a cultural icon of a bygone age. He continues to serve as part of an alarm system that awakens us when fairness and justice are under threat. Whenever and wherever his image pops up, it signals a problem. Perhaps this is the most useful message we can extract from Che's short and violent life. It may render him more valuable than the failed cause for which he fought. Perhaps for this reason the defiant image of a man clad in black beret and olive-green military fatigues continues to live in the hearts and minds of many the world over.

Before we have an ideal world—if such a world can be—we will need to come to terms with our Robin Hood mentality. Aside from its ominous connections with political upheaval and violent revolution, this mentality can take an even more frightening form. In the next chapter, we will address the enigma of how the human fairness instinct can be transformed into the most fearful variety of violence that haunts our current age: terrorism, the ultimate form of wild justice.

Chapter 10

IN THE MIND OF TERROR

It was 9:05 PM, October 23, 2002, a chilly Wednesday night in Moscow. Inside the Dubrovka Theater, two and a half miles southeast from the Kremlin, two blocks away from the Moskva River, more than eight hundred people sat raptly, watching the second half of *Nord-Ost (North-East),* a romantic musical set during the time of Stalin. Outside the theater, three vans were parked furtively under dim streetlights. Forty-one young Chechens swarmed out of the vans—twenty-two men, nineteen women, all fully armed.[1]

The Chechens rushed into the theater. When they forced the actors and actresses off the stage at gunpoint, many in the audience were enthralled by the plot, thinking it was part of the show. But the amused theatergoers were about to realize that they were becoming part of the biggest terrorist drama Russia had ever seen up to that time. Their innocent glee was replaced by extreme panic when Movsar Barayev, the titular ring leader of the Chechen squad, announced the takeover of the theater. The Chechens threatened to level the building, together with everybody inside it . . . unless the Russian military pulled out of Chechnya immediately.

While the Chechen men in military uniforms and black wool hats were busy sealing off the theater, the Chechen women, their faces veiled in black, spread out along the aisles, taking positions where they could guard the audience. The Black Widows, as they were later known, were wired together with high-power explosives strapped to their bodies. Any one of them could pull the trigger and demolish the building at any moment.

The determination of the Chechens was unyielding. "We've come to Russia's capital city," Barayev claimed, "to stop the war . . . or die here for Allah."

"We seek death more than you seek life," pronounced Ruslan Elmurzaev, the shadowy true leader of the operation, who was known by his Arabic nickname Abubakar. For the hostages, the only way to escape unscathed appeared to be intervention from Russian leadership; the government would have to meet the Chechens' demand—Russian leaders would have to act quickly by withdrawing Russian forces from Chechnya. However, giving in to the terrorists' threat was not in the Russian government's playbook. Thus, what ensued was a heart-stopping standoff. Everyone inside the theater might die at any tick of the clock.

Hostage taking by terrorists has been an unwelcome occurrence on the world stage since the 1960s. An extreme action, terrorism arrives in a diversity of guises, but revenge for perceived injustice is the most common underlying motivation.[2] "It is," according to terrorism expert Paul Wilkinson, "a useful way of inflicting hatred and vengeance on a hated enemy."[3] This was never more explicitly expressed than in Osama bin Laden's "Letter to America," "Because you attacked us and continue to attack us."[4] He was talking about America's military operations during the Gulf War, peace-keeping in Somalia, support of Israel, and protection of its interests in the Middle East centered on the oil industry. Terrorism, motivated by a desire for vengeance against perceived injustice, regardless of its legal and moral obscurity, is a form of wild justice—the same sort of revenge and retaliation we've seen in smaller self-reliant communities—writ large, often on the international stage.

Why would the Chechens resort to terrorism in the first place? Part of the answer lies in the convoluted history of the relationship between Chechnya and Russia. A mountainous region in the Northern Caucasus, Chechnya had been sparsely populated, mostly by tribal people who attempted to maintain their own ethnic and cultural identities. They had fought off foreigners, Russians in particular, since the fifteenth century. Chechnya's religious affinity with Islam increased their defiance against Russian influence. Historically, Chechens rebelled whenever Russia was weak—such as during the Russo-Turkish wars between the sixteenth and the twentieth centuries and the

Russian revolutions in the early twentieth century. Even under Stalin's iron fist, Chechens revolted against the mighty Soviet Union and resisted collectivization. But they suffered dearly for their resistance. They were dislodged from their homes in the 1940s and forced to migrate to Kazakhstan, a rough journey to the east of more than a thousand miles. Only after 1956, three years after Stalin's death, were the Chechens allowed to return to their homeland. And even then, the Russian dominance over Chechnya continued. The bitter spite between the Chechens and the Russians remained and festered.

In 1991, when the Soviet Union dissolved, the former Soviet Socialist Republics that had been forced together by the Communist regime went their own ways. Chechen separatists, sensing a rare opportunity, tried again to establish an independent nation with full sovereign rights. The Russian government under Boris Yeltsin flatly rejected their demands, because unlike the former SSRs, Chechnya had never been an independent state. Moreover, Chechnya is a strategically important region in Russia because of its oil-based economy. Were it allowed to secede, several other regions might also follow suit, using the Chechen example to declare their own independence. This domino effect was the last thing Russia wanted to see.

Despite Moscow's adamant stance, the Chechens refused to cave in. On the contrary, anti-Russian sentiment grew, and separatist activities increased. Yeltsin responded by sending military forces to regain control of the region. In the process, savage and indiscriminate Russian airstrikes killed thirty to forty thousand civilians in Grozny, the capital of Chechnya.[5] Even so, the Chechen guerrilla fighters inflicted heavy casualties on the ill-prepared and demoralized Russian forces. In the end, Yeltsin and Chechen resistance forces declared a ceasefire, followed by a formal peace treaty in 1997.

Not long after that, a widespread economic downturn swept through the vast expanse of the new Russian Federation. As it was already one of the poorest regions, Chechnya was struck particularly hard. The Chechen separatist movement, merging nationalism with religious extremism, gained new strength from the prevalent public disaffection. In August 1999, Moscow and several other cities were shaken by a series of apartment bombings, resulting in over three hundred deaths. Chechens were widely blamed for the killings. Moscow again mobilized military forces in Chechnya, and the Second

Chechen War broke out. Having learned from their previous experience, the Russians came in this time with overwhelming force and mercilessly crushed the Chechen rebels.

This brief historical account highlights a critical circumstance. Before the Dubrovka Theater attack, the tension between Russians and Chechens had been simmering for centuries. For the Chechen separatists, the recent and brutal defeat at the hands of the Russian military opened old wounds that had never really healed. Yet facing this overwhelming goliath, the Chechens could do little but resort to terrorist attacks.[6]

The Dubrovka Theater hostage crisis, despite its deep historical and cultural context, shared many features common among modern terrorist groups: al-Qaeda in the Middle East, the Irish Republican Army in Northern Ireland, the Basque separatists in Spain, the Tamil Tigers in Sri Lanka, the East Turkistan bus bombers in China, and many others. Behind all this spite is the same old human fairness instinct.

Similarities aside, human spite, when compared with TFT, bears a major difference in the way it works. TFT has no memory beyond the previous iteration with its partner. Human memories go much deeper. Language allows us to pass on our stories—including past clashes with our enemies—to future generations through legends, ballads, sagas, and folklore, various versions of what we know as "word of mouth." Even so, oral yarns tend to pale over time; furies for revenge die down.[7] As a result of the arrival of written languages, however, forgiving by forgetting is no longer an option. History inscribed in permanent media—stones, baked clay blocks, bones, bamboo strips, parchment, or paper—is an enduring reminder of our ancestors' bloody feuds. Time is no longer the best healer for wounds in human affairs.

Modern media have made historical information all the more accessible. Transmitted by radio, the Internet, and twenty-four-hour cable news, history is slickly packaged and delivered with lively graphics and video footage, often accompanied by commentary that is suggestive or even incendiary. Overwhelmed by the injustice and humiliation suffered by our ancestors, we may become less forgiving than we normally are. Hence, history serves as a compelling justification for renewed aggression. Moved by feelings of vengeance toward a hated historical enemy, even an entire nation can be drawn into a frenzy of bloody revenge.

Slobodan Milosevic, president of the former Yugoslavia, used history to fan his followers into taking revenge on the nation's Muslims. On June 28, 1989, he addressed a million Serbs in the fields of Kosovo, where, exactly six centuries before, the Serbs had been defeated by the Ottoman Turks. Legend has it that from that day on Serbian fathers would hold their newborn sons and cry, "One more avenger for Kosovo."[8] Although Milosevic's speech was couched in the terms of unity and national cooperation, the flames of nationalism flared out poignantly:

> The Kosovo heroism has been inspiring our creativity for six centuries, and has been feeding our pride and does not allow us to forget that at one time we were an army great, brave, and proud, one of the few that remained undefeated when losing. Six centuries later, now, we are being again engaged in battles and are facing battles. They are not armed battles, although such things cannot be excluded yet.[9]

The timing of Milosevic's speech, the selection of the location, and the speech's innuendos reopened a six-hundred-year-old wound dividing the Muslim Bosnians and the Eastern Orthodox Serbians. Motivated by the feelings of ancient ethnic revenge, Serbian paramilitary forces unleashed a genocidal war against the Muslims in Bosnia.

Akin to the conflicts between Russians and Chechens or between Serbians and Muslims, the 1994 genocide of the Tutsis by the Hutus in Rwanda was partially rooted in past animosities between the two ethnic groups. They had come into conflict in other African nations such as Burundi and the Congo, in addition to Rwanda. There are similarities as well with the ongoing conflict between Palestinians and Israelis, an apparently endless cycle of spite sustained by a history of clashes and bloodletting.

Many other hostilities between nations—Japan and Korea, Japan and China, China and Vietnam, China and India, India and Pakistan—can be traced deep into a history of conflict. It's one of the most regrettable patterns of history, yet, given our understanding of the human fairness instinct, one of the most reliable patterns as well. Time and again past conflicts are employed as a justification for staging new violence. Ironically, this turns the popular saying—"Those who forget history are condemned to repeat it"—on its head. Indeed, those who remember history too well are doomed to repeat it!

"I believe that the social historian," reflects Maurice Keen in delving into the motive behind the literary creation of Robin Hood, "will be best advised to view them [the Robin Hood ballads] in the context of an age plagued by lack of 'governance,' when methods of enforcing law and order were still rudimentary and justice nearly always partial."[10] Weak governance continues to be a condition that favors the emergence of revenge-motivated violence, including terrorism in the contemporary world.[11] Diplomatic negotiations, economic sanctions, or condemnations by the United Nations often lack the teeth to puncture inflating animosities between conflicting nations. To some degree blinded by their fight for justice, nations or peoples on the opposite sides of such a conflict are often so locked into bitter revenge that only military intervention from third parties can break the cycle of violence. In such situations, international affairs operate according to the code of the street, and wars break out as the level of spite escalates. It's hard to imagine how much worse the ethnic cleansing might have been in Rwanda or the former Yugoslavia had military intervention on the part of NATO or the United Nations been withheld. Likewise, the utter defeat of the Axis nations during World War II, despite massive casualties on both sides, became inevitable. These examples illustrate how fair legislation followed by firm law enforcement is a vital measure that can prevent international conflicts—including terrorism—from escalating.[12]

By taking hostages, the Chechen terrorists hoped to press Moscow into withdrawing Russian armed forces from Chechnya. The demand, of course, was hard to meet. Any military deployment is a major move. Had the Russian authorities complied immediately, it would have taken time. Forced to be realistic, Abubakar, Barayev, and company became antsy, hoping for a meaningful negotiation to get the process started. This wish, unbeknownst to the hostage takers, was not to be met. The Russian government, under the rein of Vladimir Putin, took the hostage crisis as a challenge to its authority. Putin recognized that giving in to the terrorists' demand would be perceived as a humiliation of the Russian leadership. Given the historical context of revolution, war, and hardship, there was simply no precedent for submission. During World War II, for instance, vast numbers of Soviet troops were sent to the front without weapons; waves of bare-handed soldiers charged the German Panzer divisions, picking up guns from their fallen comrades. In their protracted war in Afghanistan during the 1970s and 1980s, the Soviets, despite suffering heavy casualties, refused to admit defeat; they withdrew their troops without signing a formal peace treaty. A former head of the KGB, the security agency of the Soviet Union, Putin was accustomed to heavy-handed solutions. How likely was he to be brought to his knees by a few dozen terrorists?

Indeed, instead of meeting the Chechens' demand, Moscow was busy preparing its elite counterterrorist Special Forces—*Alfa* and *Vympel* units—for a final showdown. While the terrorists stayed put, waiting and hoping for a concession from the Russian government, the Special Forces were drilling in a similar venue in preparation for retaking control of the Dubrovka Theater.

At 7:00 PM on Friday, October 25, nearly forty-eight hours into the siege, Barayev lost his patience, announcing that they would begin to kill hostages at midnight if Moscow still had no definitive response to their demand. This precipitated the final crisis. If unable to react quickly and decisively, the Russian government would suffer a huge blow to its popularity.

Putin remained silent in public until just one hour before midnight, the deadline for hostage execution. He eventually responded by having General Viktor Kazantsev, the commander of Russian forces in Chechnya, call Barayev. Kazantsev promised to fly to Moscow and meet him by 11 AM the next day for direct negotiations. "All I ask is, don't lose your cool," Kazantsev reassured Barayev on the phone. "Don't take any unnecessary steps."

Moscow's move worked well. Kazantsev's call soothed the agitated terrorists, who chose to wait for the general's arrival. But it turned out to be only a short respite for the hostages. Putin was buying more time in preparation for storming the theater. He was planning a risky gambit in his chess game with the terrorists.

"Terrorism is not caused by poverty," states the *9/11 Commission Report*. "Many terrorists come from relatively well-off families. Yet when people lose hope, when societies break down, when countries fragment, the breeding grounds for terrorism are created. Backward economic policies and repressive political regimes slip into societies that are without hope, where ambitions and passions have no constructive outlet."[13] It is true that terrorists are desperate people. Though hopeless and helpless, they're still driven to satisfy their vengeful desires. How can we expect them to refrain from indiscriminate violence?

The most puzzling and disturbing aspect of terrorism is that its practitioners pursue vengeance for perceived injustice by targeting civilians. As such, terrorism appears paradoxical, as it seeks to serve justice by creating new injustice. This prompts us to ask: do terrorists have any moral sense, since they disrespect innocent lives? One reason for harming civilians appears to be convenience. Military and government targets are more strongly secured and harder to take; civilians provide a more vulnerable alternative for revenge and retaliation. In addition, they're more readily used as bargaining chips. With the lives of civilian hostages hanging in the balance, it is easier to force the enemy government to yield to the pressure of public opinion.[14] So, as illustrated by the Dubrovka attack, terrorism can be viewed as wild justice run amok. But there is another layer of explanation.

Compared with the violence inflicted on coworkers or random targets in the street by disaffected mass murderers, terrorists—especially political terrorists—are almost never irrational.[15] Despite the extremity of their ideology and actions, their goals are clear. Their plans are carefully laid out, and their actions are meticulously measured. They often have sympathetic supporters who provide moral justification and financial backing. If reckless killing of the innocent were to result in the loss of these supporters, it would be self-defeating for terrorists to target civilians.[16] So the relevant question—instead of why terrorists kill the innocent victims—becomes why the terrorists' supporters approve of such tactics.

This calls the very concept of innocence into question. What is innocence?

When Bernie Madoff, for instance, swindled his investors for $65 billion by cooking up a gigantic Ponzi scheme, were his sons innocent? Was his wife innocent? If the answer is yes by American legal standards, why did so many victims vent their anger toward the Madoff family members? This public rage was at least partially responsible for the death of Madoff's eldest son, Mark, who committed suicide in December 2010.[17]

When two people cannot agree on the definition of innocence, there should be a deeper reason beneath the divide. In legal and rational terms, from a Western perspective, innocent civilians are those not directly involved in the events or issues at hand. However, what appeals to our emotions may be different than rational legality. Natural selection has shaped our minds to handle an extremely important task—to discriminate between friends and enemies. We are tuned into dividing our world into us versus them, or in the lingo of psychology, ingroup versus outgroup members. Mao made this clear in a memorable statement, "Who are our enemies and who are our friends? This is the first question for revolutionaries."[18]

Making the distinction between friends and foes relies on assessing the other in terms of closeness in kinship, familiarity, gender, ethnicity, religious affiliation, cultural background, geographical proximity, to name some of our key considerations. Although these identity factors do not weigh equally in our minds, by and large, the more people have in common, the more likely they will regard each other as friends. Conversely, the less they share, the more likely they will see each other as enemies. After a person identifies friends and enemies, his or her brain response will be divided between two different regions of the medial prefrontal cortex: the ventral region for the friends and the dorsal region for the foes.[19]

But these two specific brain regions are apparently not set in stone. Friends and foes are relative notions, and they can be recatalogued, depending on situations. When two villagers vie for a favorable fishing spot, for instance, they view each other as rivals—though normally not to such a degree as to treat each other as enemies. But when planning a raid on a neighboring tribe, rivals in the same band become allies. When competition and threat loom from an outside group, uncertainty and stress can nudge people to turn toward ingroup peers and denigrate outgroup members.[20]

Innocence, too, is a relative concept. It is well illustrated in a story related by biologist David Barash and psychiatrist Judith Lipton. In an eighteenth-century Swiss farming village, a blacksmith was found guilty of murdering a farmer. The enraged villagers, despite their intense desire to punish the perpetuator, were forced to grapple with a difficult dilemma: the death of the only blacksmith would eliminate a critical service in their community. Yet one way or another, justice had to be served. In the end, they found a solution: hang a tailor instead because the village had seven of them. Eliminating one would not disrupt provision of a valuable service, and the profession of tailor was thought sufficiently close to that of blacksmith.[21] Odd though this solution may seem, the case illustrates how an otherwise completely innocent person could be considered guilty by association—association in terms of professional similarity. The tailor was sacrificed to satisfy the community's desire for retribution—in a way that was deemed relevant but didn't harm other interests.

Psychologists have long been puzzled by the heuristics of ingroup/outgroup logic. People not considered friends often become enemies for no other reason than falling outside the familiar circle of friends, though they have no tie whatsoever to real enemies. From early childhood playground games, the rule that "either you are with us or against us" is applied in situations ranging from personal relationships to international affairs. It is this "guilt by association" that gives rise to such things as stereotyping, profiling, group identity, and group boundaries, all intended to separate friends from foes. Thus, it makes sense that about a quarter of Turks, a third of Indonesians, and over half of Jordanians and Lebanese Muslims believed violence against civilians could be justified in the name of defending Islam, as revealed in a Pew Survey in 2005.[22] Likewise, people who identify themselves as Arabs tend to be more inclined to support terrorist organizations that target Western people and interests.[23] For Islamic fundamentalists, Jews and Americans are the main enemies.[24] The same rule applies to Osama bin Laden, who issued a Jihadist fatwa on February 23, 1998, calling it "the duty of all Muslims to kill US citizens—civilians or military, and their allies everywhere."[25] The same logic was used as justification for the 9/11 attacks: "America had attacked Islam; Americans are responsible for all conflicts involving Muslim. Thus Americans are blamed when Israelis fight with Palestinians, when Russians fight with Chechens, when Indians fight with Kashmiri Muslims, and when the Philippine government fights ethnic Muslims in its southern island."[26]

As terrorism is "a struggle for audience support,"[27] a grassroots desire for revenge to achieve justice can nourish a terrorist cause. Because of this, terrorists see themselves as martyrs, heroes, liberators, and holy warriors, fighting for a just cause—even though they harm civilians in the fight.[28] This phenomenon is known as moral disengagement. Our inability to agree on a universal moral standard places us in a minefield of moral relativism where one man's terrorist is another man's freedom fighter. Indeed, few terrorists vilify themselves for ruthless actions against civilians; instead, they view their civilian victims as conspirators of their enemies, and thus as surrogates that satisfy their desire for revenge. "Whether based on secular ideology or religious faith, however, belief in the absolute justice of the cause has characterized the propaganda of all terrorist organizations."[29] Again, the shadow of Robin Hood appears on the scene, imprinted in the minds of terrorists—they are simply serving justice, for those they represent, in unconventional ways. "Their violence is really part of the common human reaction to unrelieved oppression," writes Keen. "Injustice may be tolerated for a long while, but if it is sufficiently unjust, crime which is committed against the perpetrators will cease to be such. Ultimately it will acquire a halo of glory because, inhuman as it may be, it is perpetrated in the name of humanity and justice."[30]

Innocence, while a relative concept, shows some contextual nuance and flexibility nonetheless. The Chechen terrorists in the Dubrovka crisis were influenced by global norms. They wrestled with the boundary between friend and foe, partly out of concern they might lose their grassroots support, which was key to the perpetuation of their cause. Innocence, for them, was not black and white, but instead calculated in shades of gray. They were quite willing to release the Chechens, Muslims, some Russian children, and two pregnant women among their hostages. They also let go an injured man who was accidentally wounded. Additionally, they agreed to free around thirty foreigners, though this never came to pass due to special circumstances.

The Chechens' calculations of relative culpability made the fallacy of guilt by association seem almost logical. They considered all Russians enemies, regardless of the fact that none of the hostages was directly involved in the

Russian military actions in Chechnya. What mattered to them were nationality (Russian vs. Chechen), religious affiliation (Muslim vs. non-Muslim), and assumed political stand (for or against Russian military actions in Chechnya). By comparison, the Russians were more "guilty" than the unfortunate tailor in Barash and Lipton's story.

Redirecting aggression to vent the driving desire for revenge and retaliation is fairly common in the international arena. This makes wealthy Western nations particularly vulnerable to the collective envy of poor nations beleaguered by domestic troubles. As globalization continues to expand its reach, the affairs of people who once seemed irrelevant to the industrialized world have become far more important. Impoverished nations find it easier to hold foreigners up as scapegoats for domestic ills such as poverty, unemployment, illiteracy, wealth polarity, and political instability. By doing so, they can divert attention from local problems while still satisfying the desire for vengeance. Apparently, the rise of international terrorism since the 1960s has been partly linked to domestic tensions in the Middle East, especially in nations plagued by dictatorship, corruption, and incompetent governance. Western intervention has aggravated many social ills in these nations, and it is frequently cited by locals who are looking for someone to blame for their plight. Mutual suspicion, ill will, and animosity have continued to build between the developed and developing countries.

Unfortunately, for many, the United States best fits the role of scapegoat. Economic and military strength aside, it's difficult to ignore the cumulative effect of hawkish American policies in political and economic affairs in Asia, Africa, and Latin America over the past century. Widespread disaffection provides fertile ground for guerrillas and terrorist organizations to emerge and flourish with considerable grassroots support in these regions. Because of its dependence on imported oil and interest in keeping foreign supplies flowing, America is perceived as worse than simply an unfair player. "*If anything mattered more than religion for many Americans . . . it was oil* [emphasis added]," writes historian Patrick Tyler. "Access to Middle Eastern oil was essential for America's postwar development. In the event of renewed world war, the control of oil would be pivotal to Western security." He elaborates,

> Because of this, the victorious powers that emerged from World War II—Great Britain, France, the Soviet Union, and the United States—came to see the Middle East as a zone of competition. British and French colonial armies . . . gave up bases along the Suez Canal and in Algeria and Bahrain. It was not surprising that America would come to the fore. . . . But it was far from inevitable that America would prevail, and what is most striking about the century of U.S. effort is the record of vacillation, of shifting policies, broken promises, and misadventures, as if America were its own worst enemy. The Middle East would never have been an easy region to master for the purpose of protecting U.S. national interests, but our mistakes have made it progressively harder to do so. The cumulative effect of American diplomacy has fed the anti-American rhetoric that is heard on the streets in Middle Eastern capitals, even in countries nominally allied with the United States.[31]

Thus, the United States is partially responsible for current problems and instability in the Middle East. And among the most criticized of American policies are those regarding the Israeli-Palestinian relationship. By siding with Israel, America has lost numerous opportunities to serve as an honest broker to dispel the deeply entrenched spite between these adversaries. Letting the decades-long conflict continue not only hurts both Israelis and Palestinians but also brings the ire of many Middle Eastern nations to focus on America for its apparent double standard, making Americans a prime retaliatory target for terrorist attacks. "Effects on daily lives are more likely to become a root of terrorism when people feel the immediate, harsh, highly visible hand of either an occupying power or a repressive regime," security expert Paul Pillar points out. "And the United States will be a target of popular wrath if it is seen as supporting that hand. The leading example is the anger of Palestinians who feel every day the yoke of Israel, and who express their anger both at Israel and at the United States, which is widely, even though incorrectly, assumed to have a role in almost everything Israel does."[32] Although the Clinton administration helped with the negotiation of a major peace deal between Israel and Palestine, the suspicion and resentment remain too deep to allow many Middle Easterners to place their trust in Americans' good will.

In the roadmap laid out by TFT, there is rarely an easy exit for the vicious cycle of spite. This is illustrated almost perfectly by the relationship between Israel and Palestine. Currently, arriving at a substantive peace agreement seems nearly impossible. On one hand, with hatred between the two peoples steeped in decades of continual conflict, any compromise or concession from either side is likely to be seen as a sign of weakness by its own people. On the other hand, Israel's considerable military, economic, and political clout virtually eliminate the possibility of ending this vicious cycle through external intervention from the United States or other international organizations such as the United Nations and NATO. The tragic combination of internal and external forces results in an internecine game of TFT, defection answered by defection on every move, from which neither side can escape.

Will Israel and Palestine have to suffer the spite indefinitely? The answer is a cautious no. The peaceful resolution of conflict in Northern Ireland provides a precedent with a ray of hope. It was a success story of tireless international negotiations, combined with mutual restraint and concessions from both the Irish Republican Army (IRA) and the British government. At least in theory, there is no reason why Israelis and Palestinians cannot follow suit.

Optimism aside, we should not be overly sanguine, for the vicious cycle of spite may yet end in mutual destruction—as we saw so often in the previous chapter. This sad scenario played out once again in the finale of the Dubrovka hostage crisis.

At 5:30 AM on Saturday, October 26, the counterterrorist Special Forces began execution of their plan to retake the theater. They first flooded the theater's ventilation system with a secret gas. Inside the theater, a gray smoke poured down from the ceiling vents with a light hissing sound and a faint acidic odor. While those awake were quickly put to sleep by the mysterious gas, the gunmen sensed a sudden turn of events. Bewildered and confused, they rushed to open side doors, randomly firing into the streets. Meanwhile,

the Black Widows, unable to leave their posts guarding the hostages, were knocked out by the gas, some fainting while standing.

For fifty-five minutes, the gas continued to spew from the vents, until all inside the theater—hostages and terrorists—had passed out. Safe enough, the Special Forces moved in. Once inside, the commandos spread out, finding and finishing off all unconscious terrorists at point-blank range. No casualties were tallied among the hostages and the antiterrorist troops. It appeared the operation was a spectacular success—or so it seemed at that point.

Before the celebration could begin, the sleeping hostages had to be awakened. This proved to be the most daunting task of the entire crisis. Those who were involved in the rescue suddenly found themselves facing an unanticipated problem—the shortage of life-saving gear and experienced medics. The 970-some unconscious hostages required immediate treatment. Any delay might lead to death or permanent damage to their nervous systems from the toxic gas.

The severe shortage of physicians and medics forced inexperienced volunteers to come to the hostages' aid. This turned out to be disastrous. As the unconscious hostages were being carried outside, improper handling caused many to choke on their own vomit. Some, after having survived the drawn-out ordeal, died on the spot. Though there was plenty of antidote available, many victims failed to get injected with it quickly enough, as the extremely shorthanded medics were frantically trying to treat hundreds of patients at once. It was pandemonium. The dead and still living were strewn together outside the theater. Many expired on their way to the hospital in ambulances and buses. Exacerbating the situation was the fact that the Russian authorities stubbornly refused to reveal the ingredients of the mysterious gas, leaving doctors at a loss for how to treat the patients.[33] In the end, 129 innocent hostages died, according to official statistics. Many others suffered long-term disabilities from the effects of the gas.

Ironically, most of the bombs on the Black Widows turned out to be duds that could not have been detonated. Apparently, they had been sabotaged either by Russian agents who had infiltrated the Chechen terrorists' network or terrorists themselves who sought to avoid provoking more vicious retaliation from the Russians. Given this tragic ending, it's hard to avoid asking the questions: Had Moscow not attempted the rescue, would there have been so many deaths? Had the Chechens started executing hostages, would they have killed 129? An otherwise hugely successful rescue ended in a massive disaster. Who should bear the blame? Putin? The Russian government? The field personnel in charge of the operation? Or, ultimately, the Chechen terrorists who created the crisis in the first place?

Clearly, Moscow didn't anticipate a massive tragedy due to a seemingly minor glitch in the rescue mission. However, even if everything had gone Putin's way, would he have been able to eradicate Chechen terrorism through this operation?

Certainly, the high casualty count was accidental. But the case also reveals the risks and ineffectiveness underlying a heavy-handed approach to fighting terrorists. Attack begets counterattack, which again begets counter-counterattack, ad infinitum. As we've seen, there is no easy exit from the cycle of human spite. The tragic coda of the Dubrovka hostage crisis was neither the beginning nor the end in the spiral of wild justice through terrorism in Russia. Before that, Chechen terrorists had taken hostages and bombed apartment complexes in several Russian cities, killing hundreds of innocent people. Two years after the Dubrovka attack, the Chechen separatists continued their terrorist campaign, leading to another major hostage crisis in a school in North Ossetia. This ended in the deaths of 331 hostages, including dozens of children, when Russian security forces stormed the school facilities. From hostage taking to crisis resolution, this one was almost a replica of the Dubrovka tragedy, only on a larger scale.[34] Why was Putin unable to stamp out terrorism with his heavy-handed approach?

Part of the answer lies in grassroots support for terrorism. History shows that "terrorists in most cases had vast popular support from their own populations."[35] Military operations can disrupt terrorist networks, but they cannot cut off recruitment channels for terrorist organizations, nor can they discourage terrorism as a strategic option for achieving political goals and meeting the psychological need for revenge. As was made abundantly clear by the Dubrovka tragedy, military operations in response to terrorist acts have at least two major downsides. They may "provoke a wider conflict," resulting in much higher costs than the intended fatal blow on the terrorists, and civilian casualties may topple the counterterrorist cause from the moral high ground and lead to a loss of sympathy in the international community. "There is, at the strategic level, the much greater danger of military over-reaction undermining the values of the rule of law and protection of human rights, which democracies have a duty to uphold."[36]

To put it simply, fighting terrorism may backfire, as the death of every innocent can provoke public outrage and become a recruiting tool for terrorist organizations. The Chechen Black Widows in the Dubrovka attack were a superb case in point. Before joining the terrorist cause, they were ordinary people—teachers, pharmacists, accountants, actresses. Their husbands, brothers, parents, and other friends and relatives had been killed by the Russian forces. How can we expect human beings to treat the violent deaths of their beloved nonchalantly? Indeed, one study finds that three-quarters of one terrorist group were friends and relatives of other terrorists.[37] Targeted killing of Palestinian terrorists, another study shows, resulted in more suicide bombings in Israel by the Palestinians.[38] Military operations cannot subdue the fairness instinct from which wild justice flows and flourishes. "Terrorism," in this sense, "will only be defeated if we act to solve the political disputes or long-standing conflicts that generate support for it," pointed out former UN Secretary-General Kofi Annan. "If we do not, we should find ourselves acting as recruiting sergeants for the very terrorists we seek to suppress . . . . To fight terrorism, we must not only fight terrorists. We have to win hearts and minds."[39]

Terrorism will likely prove impossible to stop so long as its grassroots support remains strong. "Military responses to terrorism," writes antiterrorist expert Anthony Marsella, "will never be sufficient to contain the problem of terrorism because they do not address the root conditions that spawn it. . . . America may well defeat bin Laden and al-Qaeda, but we must recognize that others will follow. This is true because the enemy is more than a group of terrorists, it is the conditions that spawn and nurture them."[40] In this respect, George W. Bush could not have been more wrong when he addressed the nation on December 18, 2005, saying, "We do not create terrorism by fighting the terrorists."[41] He forgot to count the new recruits to terrorist organizations. His decision to fight the Iraqi war carried far more serious repercussions than simply an unfortunate mistake.[42] Who would pay for the damage it caused, especially the loss of a vast number of innocent lives? The persistent attacks by the Chechen terrorists presented Putin with the same hard truth. Counting too much on military might for a quick fix to the problem posed by terrorism, both leaders were trapped in the same dead end—the relentless evolutionary logic of TFT.

"We've got to ask, why is this man [Osama bin Laden] so popular around the world?" United States Senator Patty Murray once asked. "He's been out in these countries for decades, building schools, building roads, building infrastructure, building day-care facilities, building health-care facilities, and the people are extremely grateful. We haven't done that."[43] Realistic and practical though this comment might seem, it nevertheless generated shock and derision on the part of hawks. Characterizing Murray's remark as "jaw dropping," commentator Austin Bay taunted, "America desperately needs a new generation of defense Democrats, liberals with clear heads and sharp swords."[44] But the truth is that America has used "sharp swords" in the Middle East and elsewhere for many years. Why have terrorist activities become more deadly and widespread, spreading to even on the US soil? In the Muslim world, who is more hated than Americans? Yet, why do many still expect "sharp swords" to be the *only* effective weapon in the effort to eradicate terrorism?

"Democratic polities under threat of terrorist attack are likely to accept extreme and absolute counterterrorist ideologies," answers sociologist Neil Smelser, "because they simplify a complex situation, because they appeal emotionally to a citizenry under threat of attack, and because political authorities can shore up their support by adopting or voicing them." But he does not stop there. "Yet such ideologies, at the very least, tend to narrow the range of strategies available to deal with terrorism, and more seriously may spawn overreactions and forms of aggression that may be counterproductive from the standpoint of dealing with the terrorist activities and their potential supporters."[45]

It is not a simple matter to generate international terrorism from the human fairness instinct—but several conditions can facilitate the process.

Adding to hawkish US policies in international affairs for much of the twentieth century, globalization has put American luxury, consumerism, and the pursuit of a lavish material lifestyle—often in a culturally insensitive way—under the crosshairs of collective envy. "Simply being the big guy—on a block, or on the globe—probably also contributes to resentment mixed with envy or with suspicion of how the big guy will use his strength among some of the less powerful," agrees Pillar. "In the case of the United States, much of that influence consists of a propagation of culture that is a result not so much of US foreign policy but rather of globalization and US economic strength. MTV and Big Macs have spread throughout the world—including the Muslim world, to the chagrin of the Islamists—not because of decisions made in Washington but because America is large, rich, and creative, and as such has a disproportionate influence on what flows over the world's airwaves and trade routes."[46] "Most of all," writes Marsella, "there appears to be a resentment of the United States for its efforts to impose its commercial interest on others. The terrorists resent even as they envy, and the outcome of this collision in emotional ambivalence is hatred, anger, and fury."[47]

"A good deal of this animosity—particularly in the developing world—may stem from a perception that they have been victimized by corrupt governments, backed by powerful nations and multinational corporations that have little concern for their lives, needs, or suffering."[48] By backing incompetent, corrupt, or oppressive leaders such as Ferdinand Marcos of the Philippines, Shah Mohammad Reza Pahlavi of Iran, Suharto of Indonesia, Fulgencio Batista of Cuba, and Hosni Mubarak of Egypt,[49] the United States fanned popular hatred and resentment against Americans. The current popularity of Hamas in Palestine and Hezbollah in Lebanon demonstrates that clearly. As a result, "the image of the United States as the big devil supporting local devils figures prominently in, for example, the view of Saudi Arabia that Osama bin Laden propagates."[50]

Although most statistics do not show poverty contributes *directly* to terrorism, wealth polarity and lack of social mobility, as we have repeatedly seen, tend to make the disadvantaged angry. When the pain of perceived inequality and injustice is aggravated by a history of conflict, political propaganda, and religious fanaticism, violence can be provoked and redirected toward a scapegoat, with the United States offering the most visible and satisfying target. Indeed, a 2001 poll showed that educated Palestinians appeared to be slightly more, rather than less, inclined to support terrorism.[51] The reason seemed to be that people with more years of schooling were more likely to lose their jobs than those with less education. This explains why the educated are no less susceptible to religious extremism and to the attraction of terrorist organizations.[52]

For decades, America has been vilified as an exploiter because of its voracious appetite for oil and its image as a bullying nation that often resorts to unilateral military action in the Middle East.[53] As a result, Americans often serve as the scapegoats in many Muslim nations that are rife with domestic problems. Furthermore, historical conflict, cultural divergence, and religious differences, when commingled with collective envy, can incite nationalism and extremism against Americans and American interests. In fact, several of the terrorists responsible for the 9/11 tragedy, including its ringleader Mohammed Atta, planned to go to Chechnya to fight against the Russians before they turned to action against the United States.[54] For Americans today, Chechnya is closer than it seems, as globalization hits home. The bombing in Boston on April 15, 2013, by two Chechen immigrant brothers is a painful reminder of the terrorist threat on US soil.

Do Americans need to worry about the potential aftermath of our ill-conceived policies, especially in the Middle East? Fortunately, such anxiety is somewhat ameliorated by an unexpected savior that is rescuing America's image in parts of the Islamic world. This savior is not the mighty American military; it is the Internet, facilitated by the broad use of cell phones. Freed from propaganda and government censorship, social media websites such as Facebook and YouTube allow people to share and relay information, mobilizing protests. "As of 2011, every one of the twenty-two Arab countries faced a serious political challenge," writes journalist Robin Wright. As a result, the Arab Spring Movement arose, almost completely out of the blue, in Iran, Tunisia, Egypt, Libya, Yemen, Syria, and other countries as well. "For a decade, the outside world was so preoccupied with its 'war on terrorism' that it gave little credence to efforts among Muslims to deal with the overlapping problems—autocratic regimes and extremist movements—that fed off each other. Extremism emerged largely to challenge autocrats in countries where the opposition was outlawed, exiled, under house arrest, or executed."[55] "The quest is more about 'just values'—equitable justice for all, political participation for all, free speech for all, and rights of self-determination for all—adapted to Muslim societies."[56] Behind the quest is the same old fairness instinct that emerges from every crevice of human society across the world. Theories that tout the clash of values, cultures, and religions between the West and Muslim societies—often resorted to by pundits—appear overblown.

Since the advent of social media among the people in these nations has awakened them to the deleterious effects of their own incompetent or corrupt leaderships and serious domestic problems, the image of America as an evil goliath has been greatly softened. "The far wider Muslim world is increasingly rejecting extremism. The many forms of militancy—from the venomous Sunni creed of al Qaeda to the punitive Shiite theocracy in Iran—have proven costly, unproductive, and ultimately unappealing."[57] "Every reliable poll since 2007 shows steadily declining support for the destructive and disruptive jihadis, even in communities where politics are partly shaped by the Arab-Israeli conflict."[58] As expected, support for al-Qaeda has been on the decline in the Muslim world.[59]

As the commonly misunderstood Chinese saying "Crisis also means opportunity" goes, much can be done to harness the fairness instinct, which ultimately fuels the wild justice of terrorism. "Many conditions culminating in terrorist activity," Smelser believes, "are products of centuries of colonial and other forms of international domination, the evolution of the nation-state form, and recent conditions associated with economic, political, and cultural globalization." To ameliorate these conditions, he suggests, we should consider efforts "such as aid and investment programs aimed toward making the world's distribution of wealth more equitable . . . to reduce the long-term probability of terrorist activity."[60] While it is to no avail to worry about our historical burden of myopic foreign policies, the widespread protests rising in the Muslim world for openness, equality, and even democracy since 2011 present us a historic opportunity to transform international hostility to global peace, tolerance, and prosperity.

Part 4

EQUALITY AND DIVINE JUSTICE

Fairness is an indelible element in our spiritual, socioeconomic, and biological endeavors. In the spiritual realm, divine justice, the idea of an ultimate reckoning or comeuppance, is another extension of our fairness instinct and is commonly invoked as a means to level socioeconomic and biological inequalities in human societies. Meanwhile, reproductive fairness, a biological pursuit, is one of the most important dimensions in human equality and is often pursued under celestial justification. The intersection of these two is lucidly illustrated in the conflict within and between Christian communities over the nineteenth century Mormon practice of polygyny in the United States. Many other landmark social advances—emancipation, racial and gender equality, and even democracy itself—are also often promoted under the cloak of divine justice. A universal concept shared by all major religions, including Buddhism, Judaism, Christianity, and Islam, higher justice promises and aspires to eradicate human flaws, biases, prejudices, hierarchies, wealth polarities, and other types of inequalities.

Chapter 11

THE WAR ON POLYGYNY

On March 29 and 30, 2008, Texas Child Protective Services received two phone calls from a pregnant, sixteen-year-old girl, Sarah Jessop, who lived on the Yearning for Zionist (YFZ) Ranch, a Mormon fundamentalist community at Eldorado in western Texas. The caller claimed to be the seventh wife of a forty-nine-year-old polygynist man who had repeatedly abused her, physically and sexually. Alarmed by the allegations, Texas law enforcement officers raided the ranch on April 3, taking hundreds of people into custody, including 480 children. During the raid, investigators found a number of girls between fourteen and seventeen who fit the profile of the caller; indeed, thirty-one of those girls either had children or were pregnant.[1] Many of the underage girls appeared to be involved in arranged marriages, a suspicion later confirmed by DNA evidence.[2] Twelve men were subsequently charged with sex crimes including assault and bigamy.[3]

The YFZ raid, one of the largest of its kind in US history, resembled other events in 1935, 1944, and 1953, the last of which, "the Short Creek Raid," ended with the detainment of 263 children and 85 women from a fundamentalist Mormon community in Arizona.[4] Though the federal government lost the battle for custody of the children, the YFZ raid was a major milestone in America's long battle against polygyny and arranged marriages that many believe oppress children, restrict their freedom, and violate their rights.

The indictment of Warren Jeffs, president of the Fundamentalist Church of Jesus Christ of Latter-day Saints (FLDS) brought to light another contemporary case of polygyny. (The FLDS should not be mistaken for the mainstream Mormon Church—the Church of Jesus Christ of Latter-day Saints—in shorthand, the LDS). A polygynist accused of a litany of crimes that most Americans found repugnant, Jeffs went into hiding to avoid indictment and prosecution. Prior to his capture, he was on the FBI's Most Wanted List, with a bounty of $60,000 for information leading to his arrest. He was apprehended on August 28, 2006, in Nevada and charged with multiple counts including incest and sex with minors in both Utah and Arizona. A year later, he was convicted in Utah on two counts as an accomplice to rape for his role in forcing then-fourteen-year-old Elissa Wall to marry her cousin, nineteen-year-old Allen Steel, in 2002.[5]

In 2008, Wall recounted her ordeal in the moving autobiography *Stolen Innocence*. She was not alone. Several other women formerly associated with the FLDS came forward to tell their personal stories as polygynous wives in similar books: *Church of Lies* (2010) by Flora Jessop and *Escape* (2008) and *Triumph* (2011) by Carolyn Jessop. These narrative accounts provide an eye-opening exposé of sexual and marital abuses suffered by the women who were members of plural families in isolated FLDS communities. These accounts paint a clear picture of how polygyny victimizes women, and they serve as a contemporary call to action against what seems a barbaric practice in the context of modern civilization.

The recent raid in Texas and prosecution of Jeffs demonstrate a deep hostility toward polygyny in America. Indeed, Texas law enforcement authorities were lauded for their swift and decisive action in responding to the teenaged wife's plea for help. Later, however, the mysterious "Sarah Jessop" who had called the child protection hotline was tracked down by her cell-phone number. Rather than a blond, blue-eyed teenager trapped inside the YFZ ranch, the caller turned out to be Rozita Swinton, a thirty-three-year-old African-American woman living in Colorado Springs, unmarried and childless.[6] In their rush to come to the aid of the supposed Sarah Jessop, law enforcement officers had been duped. That they were so easily tricked speaks, in large part, to a remarkable social animosity against plural marriage. This distaste is certainly not limited to law enforcement; polygyny, according to a Gallup poll conducted in 2010, was considered immoral by over 90 percent of Americans, a figure that had changed little since 2003.[7] By contrast, Americans who disapproved of same-sex marriage dropped 10 percent, from 55 percent in 2004 to 45 percent in 2011.

This attitude toward polygyny is not unique to America. In today's world, polygyny is broadly regarded as immoral in almost all Eastern and Western societies, with the exception of some Islamic and indigenous cultures. Adrienne Davis, a Washington University Law professor, supplies a cornucopia of explanations:

> Polygamy offends a diverse array of interests: traditionalists who believe in "family values" (they suspect it to be promiscuity in disguise), mainstream Christians who resent religious fundamentalism, children's rights advocates, liberals who suspect that polygamy is a combination of parental exploitation of children and religious brainwashing, hindering realization of individual desires and will . . . , romantics invested in the companionate bond that conventional marriage is imagined to engender, and even those who argue polygamy provides a cover for a range of fraudulent behavior from welfare abuse to tax fraud. And, of course, there are those who believe polygamy is an inherently patriarchal institution that subordinates women.[8]

Polygyny, according to the modern consensus, promotes gender inequality. Supporters of women's rights blast it as an evil institution that subordinates women, limits their choices, curtails their freedoms, and virtually, if not in actuality, enslaves them to men. The UN Committee on the Elimination of Discrimination against Women went so far as to condemn polygyny as an offense against gender equality: polygynous marriage "contravenes a woman's right to equality with men, and can have such serious emotional and financial consequences for her and her dependents that such marriages ought to be discouraged and prohibited."[9] For Drake University Law Professor Maura Strassberg, polygyny "not only fails to produce critical building blocks of liberal democracy, such as autonomous individuality, robust public and private spheres, and affirmative reconciliation of individuality and social existence, but promotes a despotic state populated by subjects rather than citizens."[10]

The negative opinion shared in mainstream American society, however, does not explain why so few women in FLDS communities opt to rebel against their plural marriages. Many are quick to finger social and religious coercion reinforced by "brainwashing" as causes of their acceptance of plural marriage. But beyond our intrinsic distaste for the practice, few of us have actually questioned the seemingly unquestionable: is polygyny actually bad for women?

Throughout human history, polygyny has been broadly practiced in one form or another. In a well-known survey conducted in the 1950s, polygyny (one husband living with two or more wives) was found in 83 percent of the 854 human societies analyzed, whereas pure monogamy was found in only 16 percent. Polyandry (one wife living with two or more husbands) was the rarest, accounting for just 4 percent. A more recent reexamination confirmed this pattern, with polygyny being found in 73 percent of 1,170 societies surveyed.[11] These statistics invite the question: is polygyny not only more commonplace than monogamy, but also more "natural"?

(Strictly speaking, polygamy refers to either polygyny or polyandry. However, because polyandry is extremely rare in human societies, polygamy has been used interchangeably with polygyny in news media, books, and articles. For the sake of accuracy, polygyny and its derivatives are preferred over polygamy and its derivatives here.)

Scanning the broad evolutionary landscape, monogamy occurs in only 3 percent of mammalian species. But even this meager monogamous minority is deceptive. The 3 percent includes species that engage in sneaky sexual liaisons out of "wedlock," in other words, species that are monogamous socially, but not sexually. This has led some to argue that humans, like their close evolutionary cousins, such as chimps, bonobos, and gorillas, have a natural predilection toward polygyny. From a biological perspective, that may not be far from the truth. Evolution has left telltale signs of our polygynous past. Larger in size but shorter-lived than women, men are built to compete with other men; women enter sexual maturity earlier than men. These are features characteristic of polygyny.[12]

While our biology is only suggestive of a tendency toward polygyny, history affirms this biological pattern in human societies. Among nearly all ancient civilizations—Vedic, Chinese, Egyptian, Mesopotamian, Mesoamerican—polygyny was the norm rather than the exception. Even the often-touted egalitarian Greeks and Romans—who slighted polygyny as a barbaric practice—found pleasure in taking concubines and having sex with prostitutes and slaves, with no legal or social consequences.[13] In the Old Testament, several legendary icons—Jacob, Elkanah, David, Abraham—practiced polygyny.[14] Even Jesus, despite his preference for monogamy, was not particularly hostile toward polygyny,[15] which, before 600 CE, was not considered in direct conflict with Christianity.[16]

Apparently, it was not until the fifth century that monogamy began to gain a foothold among Christians. Augustine encouraged its practice by proclaiming it a "Roman custom." In the sixth century, the Byzantine emperor Justinian asserted that, by ancient law, husbands were not allowed to have both wives and concubines. As Christianity spread across ancient Europe and Asia Minor, monogamy took root in more and more societies. Several centuries later, Christian churches began to "suppress polygamy among Germans and Slavs at a time when the Arab conquests lent ideological support to polygamy in parts of the Mediterranean and across the Middle East." Thus, monogamy, to some degree, can be view as "a by-product of Christianization."[17] Influenced by Christianity, Judaism, which was neither as widespread nor as powerful or inclined toward polygyny as Islam, followed suit and banned polygyny around 1000 CE.[18] If there might still have been some wiggle room for alternatives prior to the Protestant Reformation, there wasn't afterward. In 1563, at the twenty-fourth session of the Council of Trent, the Vatican sanctioned monogamy as the only form of marriage. "If any one saith, that it is lawful for Christians to have several wives at the same time, and that this is not prohibited by any divine law; let him be anathema."[19]

Christians, whether Catholic, Protestant, or Orthodox, take monogamy seriously—so much so that it is among the most important canons of their religious faith. Only sporadic incidents of polygyny are found in the history of Christian communities, mostly in unusual circumstances. For instance, polygyny was briefly practiced among Anabaptists in Münster in 1535 and 1536. Such an "un-Christian institution," according to historian John Cairncross, was "undoubtedly connected with the outbreak of the Protestant Reformation in Germany."[20] Indeed, Martin Luther apparently tolerated polygyny. In a letter to Henry VIII, he wrote, "I would allow the king to take another [wife] given in accordance with the example of the patriarchs of old who had two wives at the time."[21] Another such outlier occurred in 1650. After the Thirty Years' War between the Catholics and Protestants, a regional council of Catholic Franconians in Germany proposed that laymen be allowed to marry two wives.[22] Obviously, there was a severe shortage of men as a result of the protracted bloody conflict. Even so, under virtually no circumstance could polygyny be practiced without severe condemnation by mainstream Christians.

Given that polygyny has been widely practiced in so many human societies both globally and in areas that eventually came under the influence of Christianity, why do Christians come to so stubbornly oppose it? Today, in our liberal, democratic, civil society, we value and embrace diversity in race, gender, and culture; we tolerate many behaviors formerly believed deviant, illegal, or immoral—homosexuality, premarital sex, out-of-wedlock liaisons. As these gain increasing acceptance, it seems perplexing that polygyny is still considered repugnant to the vast majority of people. Why do so many Americans continue to find the polygyny practiced by communities such as the FLDS abhorrent? Why are we unable to tolerate some diversity in marriage and family patterns? Is the belief that polygyny harms women and children the only reason for our disapproval?[23]

Presuming the answer to that last question is yes, then the push for monogamy should be expected to coincide with women becoming conscious of their social and political rights. History disagrees with this expectation. Calls for monogamy predated women's suffrage and equal rights movements in Western societies by hundreds of years if not more. In England and the United States—despite monogamous marriage having been culturally entrenched in the Anglo world for centuries—married women were still deprived of many basic rights prior to 1830 and were seen largely as extensions of their husbands. "Upon marriage, the legal rights of husband and wife were merged and subsequently exercised solely by the husband. Married women had no rights with regard to their legitimate children, they could not own property, and they could not obtain a divorce. In short, a married woman had no separate legal existence of her own."[24] If the triumph of monogamy and the social prohibitions of polygyny stemmed from women's rights movements, consonant with our contemporary beliefs, such a glaring discrepancy in the timeline makes little sense.

Given the angry reaction to short-lived Protestant polygynous communities and the canonization of monogamy at the 1563 Council of Trent, the underlying motives behind the relentless push to eliminate polygyny are shrouded in centuries of religious tension and debate. The conflict-ridden experience of the Mormons in America provides a unique opportunity to dig into the evolutionary roots of monogamy as a social institution.

Historically, Mormons lived at the fringe of American society. Though founded alongside other upstart Christian sects in upstate New York during America's second great religious awakening, there were key differences that made them stand out from their peers—such as their theology, church organization, and belief in the Book of Mormon as a newly revealed and sacred Christian text coequal with the traditionally accepted Old and New Testaments. But nothing set them apart from, and brought the vilification of, mainstream society so much as their practice of polygyny. This custom seemed so aberrant and repulsive to nineteenth-century Americans that large numbers of Christians deemed Mormonism a non-Christian cult religion. Even today, long after polygyny was outlawed by the mainstream Church of Jesus Christ of Latter-day Saints, some Americans continue to hold this bias. While Mitt Romney's eventual selection as the Republican presidential nominee marked a watershed in acceptance of Mormons by mainstream America, prior to his nomination there were concerns about his faith. In 2008, during a TV appearance, a senior citizen bluntly told Mitt Romney that he would not vote for him because of Romney's religious beliefs. As Romney's bitter experience testifies, polygyny carries a historic stigma for Mormons that has never been completely erased.

Polygyny, as a sanctioned religious institution of Mormonism, was practiced for only about four decades in the history of the Church of Jesus Christ of Latter-day Saints, between 1847 and 1890.[25] Despite this short history, it has left a thick patina that taints both the past and present image of Mormonism.

In reality, polygyny was not even part of Mormonism when Joseph Smith founded the religion in the late 1820s. Only a decade later did Smith begin to practice polygyny in secret. In 1842, Smith published *The Peace Maker* to justify polygyny's biblical roots in the Old Testament. "If any man espouse a virgin, and desires to espouse another . . . , then is he justified; he cannot commit adultery for they are given unto him," argued Smith, based on a revelation he claimed to have received directly from God. This tenet became Section 132 after 1843 with its inclusion in another of the scriptural texts of Mormonism, the Doctrine and Covenants.[26]

Smith's practices and beliefs, particularly in regard to polygyny, angered local Christians, even though his practice of polygyny was only rumored at first. The hostility toward Mormonism felt by mainstream Christians was exacerbated by marked differences in economic, political, and cultural interests. Clashes between Christians and Mormons intensified in a manner best described as spite. In 1833, Mormon settlers were evicted from Jackson County, Missouri, and the conflict came to a head several years later during the so-called Mormon War of 1838, which was followed by an exodus of Mormon communities from Nauvoo, Illinois, in 1846. Smith himself had been beaten by a mob of Ohioans in 1832 for his alleged sexual relationship with a young woman, Marinda Johnson. The "gang of indignant Ohioans—including a number of Mormons" were, according to author Jon Krakauer, "resolved to castrate Joseph so that he would be disinclined to commit such acts of depravity in the future."[27]

A further indiscretion some years later with Fanny Alger, a young maid in the house of Oliver Cowdery—another of the founders of the Mormon movement, who had been Smith's assistant in translating the Book of Mormon and is still regarded as one of the three original witnesses to the authenticity of the church—led to accusations by Cowdery that Smith had an affair with the girl. Smith took her as his first plural wife in 1833. At the time, the fact that Smith may have taken Fanny Alger as a plural wife was still largely a secret, even among the Mormons.[28] At this point, despite abundant rumors, the Mormon Church in Kirtland still maintained its monogamy principle: "Inasmuch as this church of Christ has been reproached with the crime of fornication, and polygamy, we declare that we believe that one man should have one wife, and one woman but one husband, except in time of death, when either is at liberty to marry again." This statement stayed in Section 101 of the Doctrine and Covenants up until its deletion in 1876.[29]

Smith's practice of polygyny alienated several of his close followers, including his wife Emma, who had campaigned against polygyny as late as 1844. Even Brigham Young and John Taylor, who would succeed Smith and become prophets to those followers who later emigrated into Utah and formed the modern-day Church of Jesus Christ of Latter-day Saints, hesitated when they first heard about it—though both later took plural wives.[30] Some of his followers who learned of the doctrine were less agreeable; trusted friend and second counselor William Law pleaded with Smith to drop "the doctrine of plural marriage"—but to no avail. When Smith attempted to pick up Law's wife Jane, Law, like others before him, severed ties with Smith.

In April, 1844, Law demanded that Smith tell the public about his "wicked behavior." The rift between them led to a major crisis in the Mormon community. Smith expelled Law from his church while Law responded by setting up a new sect, the Reformed Mormon Church, to draw away Smith's followers. In the end, Smith won the internal struggle, and the Mormon community survived the potentially divisive infighting.[31]

In addition to Smith's personal charisma among his followers, what swayed the community in his favor was likely the ignorance of his followers: most Mormons and even founders of the church at the time still in the dark as to the core cause of the struggle between the two leaders—the controversy over polygyny. David Whitmer, another of the three witnesses to the authenticity of Smith and the Book of Mormon, left the church and later founded his own branch, the Whitmerite Church of Christ. One of its core tenets was a condemnation of polygyny, which set it apart from other Mormon groups, such as those who would later settle in Utah.

The infighting between Smith and his opponents may have ended differently had they been fully aware of the history of sanctioned polygyny. Smith survived the crisis among his followers but not the struggle with traditional Christians in the area.[32] Having broken with Smith earlier in 1844, Law began publishing the *Nauvoo Expositor*, circulating evidence of Smith's secret teachings, including Smith's overtures to Law's wife. On June 7 of that year, acting in his capacity as a civic leader in Nauvoo, Smith ordered the destruction of the press. This served to inflame passions against Smith and his followers, and only a few weeks later, on June 24, 1844, he was arrested for it. The brewing tension that suddenly erupted into outright animosity underscored a growing fear among Christian communities that Smith was becoming increasingly tyrannical and the Mormons more powerful. At that time, Smith's Mormon army—the Nauvoo Legion—claimed to have five thousand men, nearly half as many as the entire US Army.[33] The tension between the Christians and Mormons had reached the point where some were afraid a civil war might break out.

Rather than look for a peaceful solution, vengeful Christian vigilantes acted preemptively, as "bad blood between the Saints [Mormons] and the Gentiles [Christians] who lived around them had been building for at least two years."[34] While Smith was awaiting trial in a prison at Carthage near Nauvoo, about 125 militiamen stormed the jail and killed him on June 27. The mob action marked the passing of a tipping point in the hostilities between Christians and Mormons in the area and made Smith a martyr and symbol for the disparate branches and communities of Mormons from that point forward. Although there were multiple issues underlying the hostility of mainstream Christians toward their Mormon neighbors and Smith personally, it was the events surrounding polygyny—suspected, rumored, or actual—that elicited the intense emotions that contributed to Smith's violent death.[35]

The acquittal of the nine suspects indicted for Smith's murder encouraged more violence against the Mormon minority. Hundreds of Mormon houses were plundered or burned by local anti-Mormon vigilantes. The hostility from Christian communities eventually resulted in the exodus of Mormons from the Midwest in 1846. After a great deal of squabbling about who should succeed Smith as the leader of the church, a majority of the Mormons in the area got behind Brigham Young, the head of the early Mormon Church's Quorum of the Twelve Apostles. Young, following Smith's earlier recommendation, led the Mormon pioneers across the Rocky Mountains before settling in Utah.[36] Smith's polygyny was unknown to the majority of his followers until 1852, eight years after his death, when Smith's secret of "celestial marriage" was divulged by Brigham Young to the entire Mormon community that had left

Illinois and followed him to Utah. The public announcement generated a shockwave that emanated outward from Utah across the globe. "After the sacred doctrine {of plural marriage} became known outside of Utah, a nearly hysterical barrage of condemnation rained down on the Saints from afar—a barrage that would continue unabated for half a century."[37] The practice of polygyny also alarmed followers in Europe. In Britain alone, 1,776 Mormons fled from the LDS Church within six months of Young's announcement.[38]

Mormon polygyny was no less anathematized by non-Mormons than by those who had broken away from the Mormon movement. Difficult to believe as it may be for us today, polygyny was a more roundly criticized social issue than slavery prior to the Civil War. In the Midwest, the Missouri Compromise of 1820, a political agreement permitting slave-holding, allowed slavery to continue legally in Missouri. Several progressive newspapers in the area were destroyed by Christian mobs for voicing anti-slavery views. Despite tempers running high on both sides of the issue of slavery, even at times resulting in violence, those speaking against polygyny roused no such ire; indeed, no press ever suffered similar treatment for publishing anti-polygynous opinions.[39]

Given the complexity of the position of early Mormons in America, it would be logical to ask why the practice of Mormon polygyny so angered Christian communities. Was it only a convenient excuse for locals, as some scholars have speculated, to drive out Mormons for fear of their growing economic and political power in the region?[40] Although popular distrust of Mormons had been tied to other issues while plural marriage was practiced secretly, once the practice was brought into the open in 1852 the situation became much clearer. In 1856, the Republican presidential candidate John C. Frémont paired polygyny with slavery as the "twin relics of barbarism," a campaign slogan that had been used by the Republican Party since 1854, as it vowed to banish both from US territories. The Republicans advocated containing the latter to the American South, where it could die a slow death, and abolishing the former altogether.[41]

Return to the earlier question of the connection between women's rights and polygyny. At a time when only men could vote and women had little political influence, it seems odd that a campaign would emphasize women's rights in the sparsely populated Utah Territory, where statehood had not yet been granted. Moreover, since men were the beneficiaries of polygyny, why would male voters care so much about abolishing it? At the time, the women's rights movement was still in its infancy, and the Nineteenth Amendment, which granted women full suffrage, wouldn't be ratified until over six decades later. Was the anti-polygyny movement fueled by just a few enlightened advocates for gender equality? Had this been the case, Frémont's reading of the political barometer would have been totally off the mark. As the first presidential candidate of a major party running on the antislavery platform, Frémont was already taking a substantial risk with his political agenda. What potential motivation could he have had to make such a big deal out of polygyny? These questions indicate that a hidden layer needs to be peeled off before we can make better sense of the link between polygyny and society.

The practice of polygyny by Mormons was vexing to the vast majority of Americans, despite the tiny size of the Mormon community and its remote location. Thus there was broad popular support that motivated Congress to pressure the Mormons to give up polygyny, specifically by repeatedly refusing to admit Utah to the union.[42] In 1862, Vermont congressman Justin S. Morrill introduced a bill to ban polygyny in the US territories. The bill, known as the Morrill Act, attempted to weaken the power of the Mormon Church by limiting the church's property to $50,000. That the bill was proposed in the middle of the Civil War, the most divisive period in US history, is evidence of the overwhelming strength of anti-polygyny sentiment. However, despite its importance to Congress and the American public, Lincoln quietly signed the bill into law but put its enforcement on the back burner. "You tell Brigham Young," he informed a Mormon emissary dismissively, "if he will leave me alone, I'll leave him alone."[43] Bogged down as he was in the Civil War, there was little time for polygyny, but the issue remained and festered while the nation settled its conflicts over slavery and state's rights.

Hardly had the dust of the Civil War settled when the House Judiciary Committee began to ask why the anti-polygyny law had not been enforced. The problem, however, lay not only in lack of enforcement but also in the statutory weakness of the Morrill Act. By 1880, only one polygynist had been prosecuted—George Reynolds, Brigham Young's personal secretary. Even so, the case went all the way to the US Supreme Court, as the Mormons fought fiercely for their constitutional right of religious freedom. In the landmark case *Reynolds v. the United States*, the Supreme Court ruled polygyny a crime and that the federal government had the power to restrict harmful religious practices. Maintaining that polygyny was "almost exclusively a feature of the life of Asiatic and of African people," Chief Justice Morrison Waite compared Mormon polygyny to the "uncivilized" practice of Hindu women being thrown on the funeral pyre.[44] Believed harmful to democratic societies, Mormon polygyny was lambasted with the emotionally and racially charged rhetoric of "political treason by establishing a separatist theocracy in Utah" and "social treason against the nation of White citizens."[45]

To patch up the loopholes in the Morrill Act, Illinois representative Shelby M. Cullom introduced a stronger bill against polygyny in 1870. Meanwhile, Congress began entertaining the idea of granting Utah women the right to vote, hoping they would stand up and stamp out "the horrible institution of polygamy."[46] However, three thousand Utah women signed a petition against the Cullom Bill, proclaiming that they were by no means oppressed or held in bondage. In several mass meetings in Salt Lake City, some women publicly defended their rights in plural marriage, claiming that "even the poorest women could ally themselves with worthy men."[47] These pro-polygyny voices encouraged the Utah territorial legislature (which obviously supported polygyny) to enfranchise its women; when it did so in 1869, Utah became the second US territory (after Wyoming) where women could vote. We can only imagine how disturbing it was for those who were fighting with their hearts and souls to liberate women from the shackles of polygyny.

The Cullom Bill, though passed by the House of Representatives, died on the Senate floor. Yet the legal battle against polygyny in Utah continued to heat up. Four years later, the Poland Act of 1874 was passed, giving district courts more power to handle illegal plural households. "Once the Act took effect, federal prosecutors began arresting Mormon leaders en masse."[48] As the anti-polygyny chorus grew louder in American society, Congress was swamped by petitions from all over the nation encouraging it to put an end to polygyny. The executive branch also joined the campaign with brimming enthusiasm. A succession of presidents, including Ulysses Grant, Rutherford Hayes, James Garfield, and Chester Arthur, condemned polygyny as a "barbarous system"; like their predecessor James Buchanan, they were unrelenting in using federal power to crush it.[49]

These anti-polygyny initiatives paved the way for the Edmunds Act of 1882, introduced by Vermont Republican George Edmunds. A landmark in the fight against polygyny, the Edmunds Act made polygyny punishable by five years in prison plus a large fine of $500 (well over $10,000 in today's currency). The bill also greatly increased prosecutorial powers, which were weak under the Morrill Act. It required only illegal cohabitation as evidence of polygyny, rather than evidence of a marriage ceremony. (An official ceremony was difficult to prove, as many plural families were put together by consensus or through private religious ceremonies not sanctioned by the state—practices still common among modern polygynists.) Also, it barred those who practiced or even believed in polygyny from serving on juries. Moreover, the federal government limited the power of the Mormon Church by forcing it to forfeit properties valued at more than $50,000.[50] As a result, as many as thirteen thousand polygynists were prosecuted under the Edmunds Act.[51]

The government was not finished. The Forty-Ninth Congress, in 1885–86, even considered the possibility of an amendment to the Constitution to ban polygyny.[52] Though it didn't pass, merely proposing the measure was such strong evidence of the collective distaste for polygyny among Americans that the Edmunds-Tucker Bill of 1887 was hammered out as an alternative legal bludgeon. Among many stringent measures, the new bill forced wives in plural families to testify against their husbands—a major change in the American legal system, which excused spouses in conventional marriages from testifying—and withheld federal funds for European Mormons to settle in Utah. This series of federal actions in quick succession tightened the political and legal nooses on polygyny, increasing the pressure on polygynists and their families. The combined weight of these measures eventually broke the Mormon will to resist. In 1890, the LDS Church grudgingly renounced the practice of plural marriage.[53]

During this long, fierce, and all-out war against polygyny, the United

States poured everything into the fight. At that point, the nation was far from being the political, economic, and military superpower it is today. Nevertheless, America carried the battle even beyond its borders, often disregarding political and diplomatic risks. In 1907, for example, the US government added a clause to its immigration law to exclude "persons who admit their belief in the practice of polygamy," an act that sparked strong protest from the Ottoman Empire because, under the clause, all Muslims would be excluded for their religious beliefs. The United States proceeded, regardless. From July 1908 to February 1910, for instance, 131 immigrants from Turkey, India, England, Holland, Syria, and Russia were denied entry for practicing polygyny.[54]

As the fiery uproar against polygyny heated up in America, the smoke drifted across the Pacific to China. In a kingdom with a millennia-long history of entrenched and institutionalized polygyny, hostility began to brew. Here, the practice of polygyny was a much more ancient and formidable tradition than among the Mormons in the United States.

The first to challenge the tradition was Hong Xiuquan, the leader of the Taiping (Transcendent Peace) peasant revolt in the mid-nineteenth century. Born on the first day of 1814, Hong grew up an ambitious man, eager to climb high in the Chinese social hierarchy. By 1837, however, he had failed civil service exams four times. These were fiercely competitive and painfully difficult day-long exams on Confucian ethics, which were used to recruit governmental officials from society at large. This streak of misfortune might have been due more to his failure to bribe the test officials than to his lack of talent. Hong became depressed and suffered a nervous breakdown. In a moment of desperation, he came across some Christian tracts. Finding salvation in the new religion, Hong converted from Confucianism to Christianity and claimed himself the second son of God, the younger brother of Jesus.

Hong's formal knowledge of the religion was acquired in a period of two months in 1847 with American Southern Baptist missionary Issachar Roberts, who was stationed in Guangzhou. Hong wanted to be baptized, but Roberts, believing Hong had only a superficial understanding of Christianity, declined. Caring little about the rejection, Hong proceeded to establish his own religious sect, the God Worshippers, preaching his own version of Christianity.[55]

The imperial government attempted to crack down on Hong's religious society but failed. Hong and his followers, by then twenty thousand strong, rose up on January 11, 1851. After a series of victories, they gained control of many of the southern provinces of China. In 1853, Hong's God Worshippers took Nanjing, the former center of the Ming dynasty, and declared it the capital of their new regime, the Taiping Heavenly Kingdom. Hong ordered the nationalization of major assets and suspended trading activities. Believing that all men and women should be equal, he issued edicts to ban polygyny and allow both men and women to own land. Furthermore, Hong made women eligible for the civil service exams and for holding official posts, another unprecedented social revolution. For the first time in recorded Chinese history, under Hong's new heavenly regime, men were not allowed to take concubines, and both genders were treated equally. Puzzlingly, while his version of equality went so far as to allow women to serve in the military, Hong insisted that men and women be strictly separated—even married couples.

After Hong settled in Nanjing as the Heavenly King (he once christened his kingdom God's Heaven but abandoned the name upon the objection of his lieutenants), he and several high officials in this new utopian government began to take concubines. Before he died in 1864, Hong collected as many as two thousand three hundred young women to populate his harem, fundamentally nullifying one of the most important Christian principles, especially in modern Christianity. More alarmingly, Hong revised the Bible according to his political needs and personal tastes, confirming Rev. Roberts's suspicion that Hong's understanding of Christianity was, at best, shallow.[56]

Hong committed suicide upon the impending defeat of his Heavenly Kingdom by the Qing government in 1864.[57] His attempts to transform Chinese society, to bring equality to men and women, and to challenge traditional Chinese family patterns were like a meteor, flaring brightly for a moment before plunging into darkness and oblivion. His unfinished revolution did not resume until decades later, toward the end of the nineteenth century.

The banner of social reform would eventually be picked up by the man many Chinese (especially Taiwanese) revere as the "Father of the Nation" and the driving force in bringing China into the modern world—Dr. Sun Yat-sen. Born two years after Hong died, and a physician by trade, Sun is most often remembered for leading the Chinese people in bringing down the reign of the last imperial dynasty, the Manchurian Qing. Sun instituted republicanism in China under his blueprint of the Three Principles of the People: nationalism, democracy, and the people's livelihood (welfare)—a philosophy that echoes Lincoln's view regarding the nature of democratic government, outlined in his Gettysburg Address only a few decades earlier.

Western culture, in particular Christianity, had a long and profound influence on Sun's personal development. Supported by his elder brother Mei, Sun studied for four years in Christian schools in Hawaii from age thirteen. "The instruction he received in Christian doctrine and his observance of conversion among some of his Chinese classmates fascinated him."[58] Noting his brother's interest in Christianity, Mei sent his younger brother home to China in 1883 to "take this Jesus nonsense out of him." However, this did not have the desired effect. In 1886, Sun began to study medicine under Christian missionary John Kerr in a Guangzhou hospital. Six years later, he obtained a license to practice medicine in Hong Kong, where he was baptized by American minister Charles Hager of the Congregational Church.[59]

During his years as a medical student, Sun devoured Western books on philosophy, economics, history, and science—including the Old and New Testaments and several of Karl Marx's works. The breadth and scope of Sun's reading "prepared him to develop his own ideas regarding China's needs," write historians Sidney Chang and Leonard Gordon. "By his mid-twenties, Sun Yat-sen's knowledge of modern western medicine, science, political, social and economic theories, Chinese classical writings, and Christianity made him an exceptionally knowledgeable and well-read revolutionary figure."[60] His rebellious ideas, developed while in medical school, earned him the apt nickname "Hong Xiuquan" among his friends.[61] Indeed, in a manner similar to Hong, Sun realized that the degenerative illness suffered by China under the Qing could not be cured by science and medicine. He quit his medical practice and jumped head first into revolutionary activities in the early 1890s.

Despite the influence of Western values—whether Marxist or Christian, Sun was first and foremost a Nationalist. Rather than challenge the traditional Chinese family system, he conformed to it. His first marriage was to a village woman by the name of Lu Muzhen, a marriage that was arranged by his family in 1884. They had three children together. Then, at the age of fifty, he fell in love with a twenty-three-year-old Methodist beauty from one of the most affluent families in Shanghai, Song Qingling. A graduate of Wesleyan College in Macon, Georgia, Qingling represented a new breed of Chinese women. Instead of bowing to her mother's objection to her marrying Sun, she ran away, and the two eventually consummated their marriage in Japan in 1915.[62] Because his previous marriage was never officially nullified, Sun, who personally believed in de facto divorce, was much condemned by Christians as a bigamist. "His distress with the affront of western Christians over his second marriage, due to his lack of a legal divorce, separated him from western churches in China."[63]

After a series of failed revolts in the 1890s and 1900s, Sun was a frequent exile, seeking refuge in many locations around the world: Europe, the United States, Canada, Japan, and British Malaya. During these numerous instances of banishment, Sun, modest yet eloquent, raised large sums of money to fund the revolution. On October 10, 1911, the revolutionaries finally succeeded in a military uprising in Wuchang, a metropolitan city in the heart of China. Sun was subsequently elected and, on January 1, 1912, inaugurated as provincial president of the Republic of China in Nanjing, the same city where Hong's Taiping Heavenly Kingdom fell forty-six years earlier. When the last emperor abdicated on February 12, 1912, the dynasties that had ruled China for over two millennia were buried for good. Chiang and Gordon note that "Sun's relentless conviction to carry on his revolutionary effort must have come from an inner strength based on the faith he had in himself, reinforced by his Christian belief."[64]

Aside from the controversy over his marriage, Sun approached Chinese cultural and political issues in a cautious, no-nonsense way. While standing for complete sovereignty of the nation and preserving the core values of Confucian tradition, he encouraged selective adoption of Western values and ways of life. Under this new, open environment, voices advocating the revival of China—both the nation and society—became more prominent in the press. In keeping with this atmosphere of openness, enlightened intellectuals and students from elite colleges and universities, many founded and run by Christian churches and missionaries, played a key role in bringing Western ideas to Chinese society.

On May 4, 1919, seven years after the birth of the Republic of China, students in Beijing took to the streets, calling for major political changes under the slogans of "democracy" and "science." This is known as the May Fourth Movement, the most important enlightenment movement in modern Chinese history. Polygyny practiced by wealthy and powerful men, though still legal according to the Provisional New Criminal Laws of 1912, was under attack. Many activists demanded that monogamy be the only acceptable form of marriage, as it was practiced in the West.

Sun died in 1925 and did not live to see monogamy written into law. The task was taken on by his successor, Chiang Kai-shek. Even more controversial than Sun's, Chiang's private life was far from being an exemplar of an enlightened monogamous man. He bounced through four marriages with two wives and two concubines, and there were many rumors regarding sexual escapades during his youth.[65] Partly out of love and partly for political convenience, Chiang courted the younger sister of Sun's wife, Meiling, so that he could be Sun's heir in every sense of the word. To win the approval of Meiling's mother, Chiang promised to divorce his wife and concubines, for Meiling was a "devout Methodist" who had graduated from Wellesley College near Boston.[66] Soon after the wedding, Chiang converted to Christianity and was baptized in a Methodist church in 1929. Chiang, according to historian Jay Taylor, "usually carried a Bible while travelling." He "found Christianity appealing because it stressed the conversion of moral thought to action and was consistent with the moral teachings of Confucius."[67]

Even so, Chiang remained a stout social conservative who doubted the feasibility of Western democracy in China. Though claiming to carry Sun's banner, Chiang nonetheless promoted Confucian morality under the name of the New Life Movement, while opposing the core spirit of the May Fourth Movement: Western democracy and science. Never a fan of capitalism, Chiang emphasized government control of social and economic systems. His position on marriage and family was rarely expressed—if it mattered to him at all—but as a Christian convert, he apparently accepted monogamy as part of his (and his wife's) personal faith.

Around the time of the May Fourth Movement, many youths rebelled against their traditional value system. It was a time when enlightened Chinese "had temple statues destroyed, discontinued the teaching of the Confucian Classics in schools, favored mixed education, ordered houses to be disinfected after deaths . . . and considered making monogamy compulsory."[68] Inspired by the zeitgeist of reviving China by smashing "backwater" traditions that many saw as responsible for holding the country back from its rightful place in the world, the Bill of Women's Movements was proposed and passed during the Second Conference of the Nationalist Party in January 1926, just a few months after Sun's death. The first such measure in Chinese history to oppose polygyny, the bill led in 1930 to formal legislation by the Nationalist government under Chiang, requiring monogamy. In Article 985 of the Civil Law, polygyny was formally banned. Given the Chinese political system and tradition of centralized authority, leaders typically played a major role in catalyzing social and political changes. As such, the impact of Sun's and Chiang's Christian beliefs and Christian lifestyles, though rarely examined by historians, should be viewed as significant to the triumph of monogamy in China.

If the mainstream Nationalist movement led by Sun and Chiang was slow and subtle, the then-minor Communist movement would eventually prove to be thorough and forceful in revolutionizing Chinese marriages and families. Puritanical defenders of equality, the Chinese Communists were more militant than the Nationalists in the war against polygyny. Even years before the Communists took over mainland China, monogamy was instituted in the Red Soviet Districts controlled by the Communists. In a bill published in 1934, monogamy was instituted as the only legal form of marriage. The ultimate victory of monogamy in China came after Chiang's Nationalist government was ousted from the mainland in 1949. A year later, the first Marriage Law of the People's Republic of China was enacted, solidly establishing monogamy while abolishing the practices of child brides and arranged marriages, both common in traditional China. Brothels were closed and prostitutes were forced to become workers. In the new capital city of Beijing, a massive police force executed a "Red Storm" on the evening of November 21, 1950. Hundreds of policemen swept across the new capital, arresting 454 pimps and 1,290 prostitutes in 224 brothels, wiping the city clean of the sex business overnight.

Equality between men and women was among the core elements of Communism and Marxist ideology. This doctrine was derived from French and English utopian thinkers such as Robert Owen, Charles Fourier, and Claude-Henri de Saint-Simon, who formulated their socialist theories under the strong influence of Christian values and traditions. Though famous for his assertion that religion is the "opium of the people," Karl Marx himself was not totally immune to its influence. He was born to a Christian family, as his father converted to Lutheranism from Judaism. Though an atheist, Marx's entire life was spent in Prussia, France, and England, where virtually no other marriage pattern but monogamy existed.

Believing that, in the words of Friedrich Engels, "polygamy, and occasional adultery, remain privileges of men,"[69] Communism deems monogamy the only acceptable form of marriage.[70] As such, it was the Communists, in their battle against marriage and gender inequalities, who finally stamped out polygyny in the vast land of China after a long and tortuous odyssey. During this historical transition, Western culture—Christianity in particular—was a main player and catalyst.

Today, though monogamy dominates across the globe, it must be seen as neither an inevitable development nor one that came easily without conflict. The war against polygyny took many centuries. In Latin America, the Catholic Spanish conquistadors, landing in the New World with an arrogant air of cultural superiority, forced Christianity on the indigenous people, conquering the Aztec and Incan empires and remaking them as what we now call Latin America. While plundering the riches of the New World, the conquistadors used polygyny as a pretext to justify racial discrimination and impose God's law by violence on the "inferior," "barbarous," "uncivilized" indigenous people. In a similar manner, Christian missionaries and colonialists wiped out polygyny across much of Asia and Africa, forcing the people to adopt Western religious values, lifestyles, and even languages.

The victory of monogamy is selectively summarized by Walter Scheidel: "In Japan, legislation against polygamy commenced in 1880. Polygamy was banned in Thailand in 1935, in China in 1953, for Hindus in India in 1955, and in Nepal in 1963. The main exceptions to this global trend have been the least secularized Islamic countries of the Middle East and more generally Sub-Saharan Africa. Despite the Quran's tolerance of up to four wives, countries such as Turkey (1926) and Tunisia (1956) have formally outlawed polygamy and others have imposed judicial restrictions on this practice."[71]

Africa presents the last significant outpost of polygyny in the world. The doctrine of monogamy was brought to the continent by colonial powers beginning in the seventeenth century. With the encouragement or coercion of European powers, local governments started formalizing laws and customs favoring the practice of monogamy.[72] Christian missionaries have long traveled the deserts and jungles of Africa to reach remote villages and tribes, bringing Christianity to these last bastions of "heathen" belief. They see monogamy as a serious mandate, virtually synonymous with the practice of Christianity.[73] Among these missionaries, the goal of saving women from polygyny is overt: "It is only by the law of the Gospel, incorporated in social life, that the Black woman will be delivered from the shame and slavery of polygamy and attain to the liberty of the children of God and the high dignity of the Christian wife and mother."[74] In their social crusade, missionaries preach monogamy, condemning polygyny as incompatible with Christianity. They often refuse to baptize polygynists. As a result, writes anthropologist Miriam Zeitzen, "Christianity with its emphasis on monogamy has significantly depressed the practice of polygyny. The spread of Christianity, and through it Western culture, undermines the normative basis of polygyny in African societies by removing some of the norms, beliefs, values and taboos associated with polygynous marriage."[75]

Even so, the message of monogamy continues to meet with stiff opposition in North Africa and the former Gold Coast, long a stronghold of Muslim culture. Indeed, the corridor from North Africa through the adjacent territories and into the Middle East is an area where monogamy has failed to win a decisive victory against polygyny.[76] Polygyny also continues to dominate today in Sub-Saharan nations, where the traditional lifestyle has changed little. In the Sahel region, for example, 45–55 percent of women still live in plural households. In West, Central, and East Africa, polygynous women account for 25–35 percent of the female population. In southern Africa, where the rate of polygyny is among the lowest, close to 10 percent of women remain in polygynous marriages.[77]

Despite these discouraging statistics, monogamy, even though not yet universally accepted, has become firmly rooted in many parts of Africa. Large numbers of African Christians practice monogamy and reject polygyny. It seems only a matter of time before monogamy will prevail across the African continent, as it has almost everywhere else on earth.

For centuries, the war for monogamy has been waged on all fronts—in ideological, social, religious, and legal spheres, in addition to physical battles. It appears fair to say that the prevalence of monogamy in today's world is the victory of Western culture in general and Christianity in particular. But a challenging question remains: why has monogamy triumphed in such a spectacular manner on the planet? We leave the answer to the next chapter.

Chapter 12

REPRODUCTIVE FAIRNESS AND THE TRIUMPH OF MONOGAMY

Although the rise of monogamy over the course of human history marked a sweeping transformation in families and societies, we have yet to understand why monogamy came to dominate so decisively over polygyny. Before we probe the reasons, we need to answer one question: what motivated the Christian Church to sanction monogamy? "During the late Roman and early medieval periods," according to sociologist Jack Goody, "rich widows made an important contribution to the Church, which was more likely to benefit if they remained chaste and unmarried."[1] This "could be important for the build-up of church lands, since the greater number of heirless laymen, the greater the possible benefits to the clergy."[2] In other words, banning polygyny, together with divorce and remarriage, appears to be a strategy concocted by the church to amass wealth for the clergy.[3]

Though at first glance it seems to make sense, on closer inspection, this theory doesn't hold water. True, the church could accrue more assets from its followers by encouraging smaller family size, but was it the primary cause, rather than merely a consequence, of promoting monogamy? If expanding the power of the church was the goal of the church's promotion of monogamy, but such a policy led to a reduction in the number of its followers—a consequence of monogamy—wouldn't the policy be self-defeating? Also puzzling is the fact that monogamy had not been strictly sanctioned as a social norm until the Middle Ages, despite earlier Christian advocacies dating to the time of Augustine. This does not lend support for Goody's theory.

Economists chime in with a variety of mundane reasons, irrelevant to the church. They note that polygyny, while on the decline in all developed nations today, may still be prevalent in poor countries. Since industrialization and economic development narrow down income disparity among men, especially between landholders and the landless, they think it could lead to the decline of polygyny.[4] "Monogamy emerges in advance[d] economies," one theory contends, "because of the increasing value of high quality women in the marriage market, which stems from the increasing value of their input in the production of child quality. In other words, male inequality generates polygyny, but female inequality reduces it."[5] Another theory offers a supply-side story for the upsurge of monogamy: rising labor costs and more options for women increase the cost of polygyny for men. Simply put, though men still have the same desire for more wives in industrial nations, women become more expensive to possess and maintain, and children become more costly to raise. Also, policies favoring women's education, entry into the labor force, and equal inheritance rights are conducive to monogamy.[6]

Though these economic theories based on market incentives seem to make sense, they face significant problems. First, they appear to overlook the obvious fact that monogamy is not new; it has coexisted with polygyny in all polygynous societies, regardless of industrialization and economic development. Also, they fail to explain why a wealthy man would benefit from having only one expensive wife of "high quality" instead of several affordable wives of "lower quality," whatever "quality" means in this case. Moreover, in a wealthy society—say America—where no woman is too expensive for a man as rich as Bill Gates, why would he willingly forgo the opportunity to take several more wives, if allowed?[7] What is missing in these market-based economic theories?

It turns out that, rather than *Homo economicus*—"economic man"—we are, first and foremost, *Homo sapiens*, with innate biological drives that have been evolving for millions of years. No wonder that economic models, when overly divorced from our biological nature, can lead us astray. We should not forget that humans, though endowed with the capability of thinking and reasoning, are animals in a biological world.

Human marriage, when stripped down to its core, is a biological trait shaped around our drive to reproduce. A biological principle proposed by geneticist A. J. Bateman illustrates how mating strategies affect males and females differently. As early as 1948, he found that reproductive success is limited by the amount of food available for females, whereas, for males, it depends on how many females they can mate with. Despite the focus of sex, the finding was far from titillating, as it was discovered in fruit flies—a popular subject for genetic and evolutionary investigations—and was far too remote from human sexuality for us to make an erotic connection. Bateman's finding, however, holds up astoundingly well in a wide range of animals from insects to elephants and whales. It has thus been christened Bateman's rule: females follow resources; males follow females.

Humans fall within the purview of this rule as well. Producing only about four hundred eggs in her lifetime, a woman's reproductive potential is limited. Men have an incredible edge over women on this account. With each ejaculate containing potentially hundreds of millions of sperm, a man is, *in theory*, able to fertilize all the eggs from all the women in the world. So when polygyny is allowed to proceed with no restraint, reproductive success quickly becomes skewed to favor rich and powerful men who have the resources to afford numerous women and raise the large number of children they produce. *The Guinness Book of World Records* offers a popular illustration. With sixty-nine children, the wife of eighteenth-century Russian peasant Feodor Vassilyev is crowned the most prolific mother, who somehow became specialized in multiple births (four sets of quadruplets, seven sets of triplets, and sixteen sets of twins). This staggering feat of fertility pales when compared with Mulai Ismail, King of Morocco, who is said to have sired over eight hundred children.

"Although anthropologists have used a lot of ink and paper to describe cultural reasons for the widespread permissibility of harem building," sighs anthropologist Helen Fisher, "it can be explained by a simple principle of nature: polygyny has tremendous genetic payoffs for men."[8] More precisely, as revealed by the polygynous Kipsigis in East Africa, the addition of each wife can bring a man on average 6.6 more offspring who will reach five years or older.[9] "Natural selection has designed males to strive for as many mating partners as possible, and they use whatever means, power, and resources they have to achieve this," writes economist Nils-Petter Lagerlöf. "In all societies—whether they tolerate explicitly polygynous marriage or not—it is the richest and most powerful high-status men who have more sex partners. As a corollary, more inequality in income and power tends to come with more inequality in the distribution of women, that is, more polygynous mating."[10]

Indeed, all ancient civilizations, including those in Mesopotamia, Egypt, India, China, South America, and Mesoamerica, converged in two linked aspects: despotism and extreme reproductive inequality. Men in the ruling class almost always had numerous female partners—wives, concubines, mistresses.[11] "Power is—as it always was—a means to sex," concludes anthropologist Laura Betzig.[12] No wonder ancient emperors and kings were known for keeping harems. In the legendary Ashanti Empire of West Africa, a king could have 3,333 wives. The Incan Empire was meticulous regarding the number of wives a man could own according to his status: a king or emperor could own thousands; a leader of a million people, thirty; a chief of a thousand people, fifteen; a village leader, seven.[13] In Ming China (1368–1644), the law permitted up to ten concubines for a duke, four for a royal man, three for a general, two for a mid-level official, and one for a commoner who was still childless at the age of forty.

Though less extravagant and well-defined, a similar pattern can be found in ancient Greece and Rome, and in medieval Europe. Rich and powerful men, unburdened by nominal monogamy, often sired children with mistresses, concubines, or slaves.[14] "Lothair of France, Pepin, Charlemagne and the emperor Barbarossa all had several wives," points out historian John Cairncross.[15] In medieval England, the rich and noble were "often to be found leaving bequests to bastard children in their wills," writes historian Lawrence Stone. "In practice, if not in theory," he continues, "the early-sixteenth century nobility was a polygamous society."[16]

Disparity in access to women is still common today, even in countries where polygyny is deemed immoral and illegal. In some African regions, for example, wealthy men dance around the Christian tenets so that they can be both Christians and polygynists at the same time.[17] In China, it is not uncommon for rich and powerful men to practice de facto polygyny, keeping mistresses and concubines in separate households. And in Los Angeles and San

Francisco, for instance, some residential areas are populated by such expatriate liaisons of rich Chinese men. They are often dubbed as "Second-Wife Villages" among the Chinese.

As demonstrated in many societies, this reproductive disparity confers a much larger variation in fitness on men than on women.[18] From an evolutionary perspective, men and women represent two distinct reproductive strategies. If we compare reproduction to investment, investing in men is like speculating on the volatile stocks of tech upstarts: a handful of them will make a killing whereas a large majority will go bankrupt. (How many dot-com companies from the late 1990s have become dot-gone?) Investing in women, on the other hand, can be likened to putting money in CDs or bonds. The return is stable but never spectacular. You will neither make a killing nor easily go belly-up. In a nutshell, men stand for opportunity whereas women bring stability and security to the evolutionary mill of competition for reproductive success.

The cold hard fact is that, except for rare scenarios where a large number of men are killed in wars, men and women are *on average* roughly equal in terms of reproductive fitness. Though any single man has the potential to produce far more offspring than any single woman, high-achieving men necessarily reduce the reproductive success of other men by monopolizing the pool of available women. Hence, although ancient monarchs, like successful tech companies, epitomize the fitness extremes that men can achieve, they are outliers. The vast majority of men, unlike their rulers, had no easy access to women. "The concentration of women at the top of social hierarchies," writes anthropologist Miriam Zeitzen, "implies a relative deprivation at the bottom, resulting in differential reproduction."[19]

For any stable society, marriage is a zero-sum game: one man's gain is another man's loss. Common sense tells us that when some men marry two or more women other men will have to suck up a statistical shortage of potential mates. This phenomenon is known as mate competition, a major component in Darwin's theory of sexual selection. In any given society, as a result, only about 5–10 percent of men practice polygyny.[20] The rest either are monogamous or remain celibate. Unsurprisingly, despite the widespread practice of polygyny in ancient China, monogamy was still the most common marriage pattern. The reason had more to do with economics than psychology: only a minority of men could afford concubines. A case study at Tongcheng in Anhui province shows a stunning contrast between the haves and the have-nots during the period between 1520 and 1661. Wealthy men represented only a tiny proportion of the overall male population. Of these wealthy men, nearly half held concubines, but only 4.6 percent of the peasant majority had this luxury.[21] Likewise, a 1986 survey in thirteen Arab nations revealed that only 2–12 percent of men were polygynous.[22] Among pioneer Mormons, "even when the practice [of polygyny] was at its most prevalent, no more than 20 percent of the Church's membership practiced polygamy," writes Shayna Sigman, a law professor at the University of Minnesota. Furthermore, "two-thirds of the men practicing polygamy only had two wives and less than 10 percent had four or more wives."[23]

Facing stiff competition from rich men, who usually have power to bend political, social, and legal systems in their favor, poor men often have to face dim marriage prospects.[24] Behavioral economist Gary Becker, a Nobel laureate, believes that, instead of stereotyping women as the oppressed and men as the oppressor, women may actually be better off if polygyny is allowed.[25] Since polygyny provides an additional option for women in the marriage market, women are not its primary victims, assuming no legal issues such as statutory rape or underage marriage arise. Journalist Robert Wright provides a lively and eloquent account:

> If we . . . hypothetically accept the Darwinian view that men (consciously or unconsciously) want as many sex-providing and child-making machines as they can comfortably afford, and women (consciously or unconsciously) want to maximize the resources available to their children—then we may have the key to explaining why monogamy is with us today: whereas a polygynous society is often depicted as something men would love and women would hate, there is really no natural consensus on the matter within either sex. Obviously, women who are married to a poor man and would rather have half of a rich one aren't well served by the institution of monogamy. And, obviously, the poor husband they would gladly desert wouldn't be well served by polygyny.[26]

Echoing Bateman's rule, Wright focuses on men in the entire society rather than those high up in the hierarchy, such as kings and emperors. Yet this shift in perspective reveals the entire vista of the human reproductive landscape, from the spotlight on a few high achievers at the top to the panorama of the whole society.

Since women, compared with men, put higher value on good financial prospects of potential mates—more so than physical attractiveness[27]—polygyny grants a woman the choice between being the only wife of a poor man and being the plural wife of a richer and more powerful man, assuming that her freedom of choice is not encroached upon. When monogamy becomes mandatory, however, the door of this choice is shut. For this reason, economist and jurist Richard Posner deems polygyny perhaps "the unambiguously best regime for women because it would expand their choice set." "The prohibition of polygamous marriage may appear to make no sense from the standpoint of protecting women," he states, in supporting this position. It "withdraws one option from a woman, namely that of being a nonexclusive wife. By doing so it reduces competition among men for women and thus reduces the explicit or implicit price that a woman can demand in exchange for becoming a wife—even a sole wife."[28]

By and large, resource equalities among men favor monogamy, whereas resource inequalities favor polygyny, a pattern that is perpetuated through female choice.[29] In a society where monogamy is imposed, however, cases of domestic conflict increase—for example, marital dissatisfaction, spousal abuse, and divorce—especially in families where the wife is better educated than the husband.[30] Serial monogamy, meanwhile, may also rise,[31] mainly as a consequence of spousal displeasure on the part of the wife. Believing "polygyny may be . . . a function of female mate choice," historian Walter Scheidel elaborates:

> Economists have long argued that polygyny is beneficial to most women if there is substantial inequality among men in terms of resources or other properties that are relevant to reproductive success. Simply put, a woman may be better off sharing a resource-rich husband with other women than to monopolize access to a resource-poor husband. In this context, moreover, polygyny not only benefits multiply married women but also monogamously married women in the same population by allowing them to avoid

> unions with the least desirable males. Conversely, this custom benefits male polygynists but harms other men to varying degrees, the more so the more unequally resources are distributed and this inequality is correlated with polygynous preferences. Hence polygyny tends to reinforce male inequality by matching reproductive inequality with resource inequality.[32]

Indeed, among traditional societies in East Africa, "only 10 (or 11) out of 23 studies show lower fertility among polygynously married women, and in only 4 (or 5) of these cases is the difference significant (or of considerable magnitude)." Thus, anthropologist Monique Borgerhoff Mulder concludes, "polygyny is not uniformly costly to women." So the majority of women in African polygynous societies such as those in Kenya and the Ivory Coast tend to prefer polygyny.[33]

Hark back to Utah in the late nineteenth century. It makes sense that so many women supported the practice of plural marriages. Even today, it is possible to cherry-pick its positive aspects—such as improving the career prospects of single women who have a hard time finding committed single men.[34] Luci Malin, vice chairperson of Utah's National Organization for Women, once cheerfully commented that polygyny "seems like a pretty good idea for professional women, who can proceed with their careers and have someone at home they can trust to watch their children," and Elizabeth Joseph, a polygynist lawyer, calls plural marriage "the ultimate feminist lifestyle."[35]

Powerful and wealthy men, of course, never gain in fitness from institutionalized monogamy, since the opportunities for additional wives are gone—and, along with them, the quintessential gauge of reproductive success, more children. Therefore, the biggest winners under monogamy are poor men who sit low on the hierarchy ladder. This explains why men, rather than women, tend to complain more vociferously against polygyny in human societies,[36] as illustrated in the previous chapter.

Polygyny leads to an imbalance between men and women in the marriage market and thus intensifies competition among men. A common consequence of male mate competition is delayed marriage for many men. "Scarcity of marriageable men," points out psychologist Nigel Barber, "is often exaggerated in polygynous societies where the age of marriage for men can be delayed for a decade after the marriage of women."[37]

The imbalance can be partially alleviated by tapping into younger women for marriage, a situation seen in virtually all polygynous societies, where women tend to marry young to older men who have amassed some wealth and power.[38] Statistics show that women in polygynous nations marry 4.5 years earlier and have 2.2 more children than women in monogamous countries. In Sub-Saharan Africa, where polygyny is still prevalent and twenty countries have polygyny rates over 10 percent, with Cameroon standing out with a rate over 50 percent, husbands are on average 7.0 years older than their wives, doubling the age gap found in monogamous countries.[39] In general, according to Zeitzen, "the frequency of polygyny appears to be positively associated with early initiation of sexual and reproductive activity, early and universal female marriage and minimal interruption of marriage through rapid remarriage after divorce or death of a spouse. These practices ensure that there are enough women to make polygyny possible."[40]

Intense mate competition among men can lead to an extreme paucity of single women.[41] To cope with this pressure, several rare marriage patterns emerged in traditional China. Among them was the "pawn marriage," practiced between the thirteenth and the early twentieth century. Unlike ordinary marriage, pawn marriage allowed a poor man, who was unable to afford a brideprice for a normal marriage, to obtain a woman by contract for a certain period, typically a widow. Thus, it had the flavor of a "rent-a-wife" business. The length of the term was typically three to five years, or until the woman under the contract gave birth to a child, preferably a son, for the man. In return, the man paid all the woman's living expenses.

Another Chinese adaptation to intense mate competition among men was the child bride. Practiced for hundreds of years, up until the early part of the twentieth century, in this type of marriage, a family would preemptively secure a child bride for their young son by paying a fee to the girl's family. The girl would then live in her future husband's household until both were old enough to be formally married, usually when they reached their teens. Such marriages were quite common, especially in southern China. A survey conducted in Taiwan in the 1930s revealed that 17 percent of unmarried women were child brides awaiting consummation, and a whopping 49 percent of women had been married through this arrangement.[42]

While we are fascinated by the historical Mormon leaders who lived with several wives, we rarely question how the ordinary Mormons dealt with the resulting shortage of available women. Apparently, one solution was to recruit women from outside the group. While this occurred in some cases, it was not significant in solving the problem, since studies indicate that sex ratios in polygynous Mormon communities were roughly equal. So new female recruits didn't resolve the issue of intense mate competition among men.

Another solution was to make younger women eligible for marriage. This was the way the LDS Church solved the problem historically, and it's the practice of the FLDS Church today. Studies in contemporary FLDS communities show that as many as 60 percent of women married when they were still teenagers, some as young as thirteen. Since in small communities the supply of even younger women was limited, incestuous marriage was not uncommon, leading to higher rates of birth defects, mortality, and other health problems for mothers and infants.[43] These data reflect an acute shortage of available women, a dire situation that might require some FLDS communities to "recycle" women, in a sense. In 2004, for example, Warren Jeffs reportedly banished twenty high-ranking men and reassigned their wives to others he considered more worthy.[44]

To handle an excess of males, the FLDS Church also expelled teenage boys—known as "Lost Boys"—using various excuses, some as minor as having a girlfriend, watching TV or movies, listening to CDs, or wearing short sleeves.[45] During the Jeffs reign, a large number of such boys, estimated to be over a thousand, were thrown out of FLDS communities and left on streets or along highways. The parents were forced to sever relationships with their sons. If they attempted to see or help their banished sons, they themselves might suffer serious penalties, including eviction from their homes, which were owned by the church. Growing up in isolation, illiterate, and lacking basic skills to make a living in modern society, the Lost Boys often joined the homeless and resorted to abuse of alcohol, cigarettes, and drugs.[46] The psychological trauma that resulted from this practice was devastatingly clear.

In 2006, six such Lost Boys filed a lawsuit, accusing Jeffs and the FLDS Church of expelling them to free up women for older, more powerful men, thereby reducing the competition for women. The case was settled out of court with the plaintiffs receiving $250,000 for housing and education, with part of the fund being dedicated to assisting other Lost Boys.[47] This is probably the only lawsuit in US history against the unfair practice of mate competition.

Observing such disturbing collateral outcomes of the practice of polygyny among Mormons, both historically and in today's FLDS communities, reveals a wealth of information about mate competition in humans. Finally, we have identified the most obvious yet long neglected victims of polygyny: the poor, low-ranking men, who are most likely to be deprived of reproductive opportunities.

This gives us the opportunity to revisit the question left unanswered in the previous chapter, the question of why Frémont and his Republican Party argued so vehemently against polygyny in the 1850s. Apparently, it was both visionary and self-serving. Abolition of polygyny, in addition to promoting a core Christian value, would appeal to the vast majority of ordinary American men. In doing so, Frémont was courting the bulk of eligible voters. Although he eventually finished second behind James Buchanan in the presidential race, his bold opposition to polygyny lived on. Buchanan, after being elected president, took up the anti-polygyny banner and campaigned as militantly as Frémont had against the Mormon practice, if not more so. Angered by the tales of polygyny, he even sent a military force to Utah under the guise of suppressing disloyalty.

Are women better off under polygyny? This is a more difficult and complex question to answer because the outcomes for individual women and the societies they live in may be different. Individually, women can in general be slightly better off in terms of reproductive success, especially in comparison with low-status men. Collectively, however, inequalities in reproduction, wealth, and social status associated with polygyny feed off one another, leading to a markedly less prosperous society at large. A study conducted by Stanford University economist Michèle Tertilt in Sub-Saharan Africa, one of the poorest regions in the world today, shows that banning polygyny can reduce fertility by 40 percent, increase the savings rate by 70 percent, and raise per capita economic output by 170 percent. She explains why:

> Polygyny drives up the demand for wives, which increases the equilibrium brideprice so that men have to pay a high price to marry. In equilibrium, widespread polygyny is only sustainable if population growth is high and men marry women younger than themselves. Buying wives and selling daughters becomes a good investment strategy that helps provide for retirement. This behavior can crowd out investment in alternative assets. Therefore, investment in physical assets is low, and hence the aggregate capital stock is low. Consequently, GDP per capita is also low.[48]

Therefore, the marital option for individual women in polygyny comes with an enormous society-wide economic cost, not to mention moral considerations related to women's rights. Polygynous nations are not only often mired in poverty; they are also commonly plagued by social problems and political instability. This raises the concern that, if allowed, polygyny may ultimately lead to the collapse of modern democracy. Does such a concern have any merit?

All existing organisms, including humans, are points in an unbroken genetic legacy, connected to deep time through the nexus of reproduction. From this evolutionary perspective, reproduction is the most vital—and often awe-inspiringly dramatic—event for all organisms: bull elephant seals fight to the death for the right to mate with females; peacocks grow extravagant (and otherwise useless) tail feathers to impress peahens; male praying mantises become the dinner of their mates. The very fact that we exist today stems from the fact that we are all the descendants of winners in the game of evolution—our ancestors—because they are the ones who succeeded in passing down their genes. In many ways, our body shape, our physiology, our biochemistry, and our behavior and psychology have been chiseled out by the powerful hand of sexual selection via mate competition.

Not surprisingly, women are among the most prized spoils of war. From the ancient expeditions of Genghis Khan's Mongol hordes to contemporary Yanomamö raiding sorties in the jungles of Brazil, much of human violence—looting, raping, maiming, killing—is propagated for passing down male genes. In modern warfare, the rape of civilian women continues to pose a major problem for even the most disciplined military forces. When citizen soldiers or mercenaries are under little supervision, horrific incidents of mass and wanton rape can happen.

Tracing the origins of genes in human gene pools in France, Africa, China, and the South Pacific, scientists have unveiled the telltale fingerprints of mate competition among men in history: more copies of fewer genes are from male lineages than from female lineages. This demonstrates that, in each generation in these societies—and likely a large majority of others—fewer men than women reproduce successfully. Considering the inequality in mating opportunities, rich and powerful men tend to pass down more of their genes than their poor and weaker counterparts.[49]

Successful reproduction is so important that for millennia in traditional China the happiest state was considered to be a houseful of sons and grandsons with four generations living under one roof. Childless couples, on the contrary, have been considered accursed in China, India, and many other cultures since time immemorial. In the same vein, one of the most memorable verses in the Bible reads: "Then God blessed Noah and his sons, saying to them, 'Be fruitful and increase in number and fill the earth.'"[50] Conversely, the Bible considers failed reproduction the most deadly of weapons against enemies: "Record this man as if childless, a man who will not prosper in his lifetime, for none of his offspring will prosper, none will sit on the throne of David or rule any more in Judah."[51]

Today, forgoing reproduction is generally accepted as a lifestyle choice in Western nations. Even so, being single and childless may still be a major cause of anxiety, especially in communities where familial and societal pressure for marriage and children is high. Childless couples may be the target of negative remarks from others. Their likelihood of divorce is twice that of couples with two children and ten times that of couples with four or more children.[52] Only a few decades ago, rural Irish men who failed to have a wife were more likely to be alcoholic or suffer a mental breakdown.[53] The situation can become much worse in poor, non-Western societies. Childless couples often suffer a range of excruciating consequences, including social stigmatization, marriage instability, harassment from in-laws, ridicule, loss of status, and isolation.[54]

Attitudes toward reproductive equality often motivate our social ideology and activism. Even privileged people have such a mentality. Malthus, apparently alarmed by the high fertility rate of his time, especially among the poor, issued his famous warning: "Population tends to increases in a geometrical ratio, food in not more than an arithmetical [ratio]." Instead of facing the miserable consequences of "diseases, epidemics, war, plague, famine," one solution he offered was to install "preventative checks," "the restraint from marriage" through voluntary "moral restraint."[55] Even more blatantly than Malthus, Herbert Spencer advocated for "the survival of the fittest" almost a decade before the publication of Darwin's *On the Origin of Species*. "A society would improve," Spencer wrote, "if the government did not interfere with the survival of those individuals who made the fittest adjustment to society in each generation."[56] Spencer objected to government programs aimed at helping the poor. In a sense, this marked the birth of Social Darwinism or, more precisely, Spencerism, which encouraged the eugenics movement in the United States and many European nations. The eugenics movement culminated, among several less horrific though equally misguided initiatives, in the Nazi's genocidal atrocities.

We learned in chapter 2 that one of the most important factors for the evolution of fairness is a bottom-up force that puts pressure on higher-ups to compromise in resource distribution. Since reproduction is ultimately the most critical factor in fitness, reproductive equality should have its corresponding role in the evolution of our fairness instinct. This explains why reproductive fairness can be a visceral issue even in egalitarian, democratic societies. The Hutterites in North America provide a good illustration.

The Hutterites, a branch of the Austrian Anabaptists, live a communal life. In the late nineteenth century, slightly over a thousand Hutterites settled in Montana and the Dakota Territory. But several states passed laws against them, ostensibly because of their belief in pacifism. The underlying reason, however, appears to have been their high fertility and economic prowess associated with land acquisition. This became apparent after the Hutterites were forced to move to Manitoba, Saskatchewan, and Alberta in Canada. Local governments were alarmed by the Hutterites' high fertility—higher than any other group in North America—and their practice of buying large tracts of land. Alberta passed the Communal Properties Act of 1942 to limit their expansion.

The desire for reproductive equality can be more subtle, yet no less forceful, in society today. People with fewer children may feel unease when encountering their friends, colleagues, or neighbors who have more.[57] Such sentiment can manifest in subtle ways. Look at the following example in the HBO program *Real Time with Bill Maher*:

> New Rule: Now that *19 Kids and Counting*'s Michelle Duggar is pregnant with her twentieth child, the Duggar family must move into a shoe. And someone has to tell Mrs. Duggar, it's a vagina, not a water slide. Admit it, if you've been constantly pregnant for the past two decades, it's not the children you love, it's the epidural.[58]

Behind the laughter was a sense of unfairness, which became even more evident when the news broke that her twentieth child was spontaneously aborted. The public response—instead of being sympathetic—was overwhelmingly negative, ranging from sarcasm to outright hostility. Here are a few unabridged and unedited short comments taken from the comment streams of online articles covering the spontaneous abortion:

> *God to Duggars—"TIME TO ADOPT"*
> *The real tragedy would be if they actually had a 20th child. The universe is begging them to stop breeding. Funeral, indeed.*
> *Good grief, keep your leg's closed!*
> *sorry - this is what was meant to be - i mean you no harm but you need to stop honey.*

If the Duggars can arouse such strong emotion, it is understandable that polygyny provokes even worse. Large numbers of children aside, polygyny engenders the image of entrenched poverty, poorly educated children, tax cheating, and dependence on welfare in America today.[59] FLDS communities are frequently portrayed as groups of people who deliberately "bleed the beast"—the federal government—by taking advantage of taxpayers' money. Among Americans, these reasons legitimize a sense of unfairness and injustice against polygynists. Yet outrage toward the reproductive inequality that polygyny represents—the deeper instinctual resentment—is rarely articulated.

Even serial monogamy can provoke a feeling of unfairness. Scheidel notes that "ease of divorce underwrote a degree of effective polygyny: while men were unable to have more than one wife at a time they could marry several in a row, thereby raising reproductive inequality overall."[60] Several public American personalities have been criticized for multiple marriages, including Johnny Carson, Newt Gingrich, Larry King, Donald Trump, and Rush Limbaugh. Indeed, if serial monogamy and polygyny bear a fundamental similarity, it is that they both allow rich and powerful men to have more reproductive opportunities.

The desire for reproductive fairness is no less intense in China, where, since the 1980s, the one-child (per couple) policy has been relentlessly enforced. As the central government ties the performance of officials to the execution of the policy, local governments are motivated to carry it out in a way that leaves no stone unturned. As a result, those who violate the policy may receive harsh punishment, including large fines, loss of social benefits, being fired from their jobs, or, in the case of villagers who receive few government benefits, being forced to undergo abortion or sterilization. These cases have been well-reported in Western media. Since the late 1990s, the elites—the wealthy and the famous, such as movie stars—often disregard the policy and give birth to one or more extra children. This leads to a society-wide outcry and demands for reproductive equality. This popular Robin Hood sentiment has been a little-noticed social force behind the perpetuation of the one-child policy, which, with public surveillance, has become something like a self-fulfilling prophecy.

The instinctual desire for reproductive fairness, though rarely discussed in public, manifests itself in our values, mores, conventions, and social consciousness. In this respect, other cultures are in concurrence with Judeo-Christian tradition. In many Islamic nations, where polygyny is not only allowed but encouraged, the number of wives a man can have is limited to four.[61] Even so, only a tiny fraction of men practice polygyny. For instance, polygynous husbands account for only 3.8 percent in Jordan and 1.0 percent in Iran.[62] And in Egypt, Turkey, and Tunisia, polygyny is altogether outlawed. Today, the fabled carnal excesses of Ottoman sultans have virtually disappeared.

Reproductive equality is also reflected in the fact that sexual relationships are often carefully regulated by law or religious credos. The fact that monogamy is more commonly seen in egalitarian societies than in those that are economically stratified seems to indicate that when ordinary people can voice their opinions, monogamy—that is, equality in reproductive opportunity—tends to prevail.[63] Political scientist Tom Flanagan believes polygyny is the reason that ancient nations like China and Japan failed to develop the ideas of individual rights and equality before the law. In fact, wealthy men and the ruling class imposed harsh laws to prevent crimes by men of lower strata in order to protect their reproductive advantage. And wars were often waged to obtain women beyond borders.[64]

Reproductive inequality induced by polygyny can disrupt social harmony and shake the foundation of democracy.[65] "In the context of male resource inequality," Scheidel writes, "polygyny tends to favor many women and disadvantage many men. This situation, in turn, is inherently conducive to intermale conflict and competition and thereby impedes cooperation." "In the most basic terms, reducing reproductive inequality is thought to promote collective action, which must be considered a vital element of successful state formation."[66] Indeed, data amassed from 156 states worldwide have shown that monogamous societies were significantly less corrupt, less likely to invoke the death penalty, more populous, more affluent, and more democratic when compared with polygynous societies.[67] Without socially imposed monogamy, unifying a large nation appears to be problematic.[68]

With a clear understanding of reproductive fairness and its essential role in shaping the evolution of our fairness instinct, we are ready to tackle the challenge raised at the beginning of the chapter: why monogamy has won out over polygyny across the world. We've observed that intensified mate competition among men, exacerbated by polarizing economic inequality, is a primary cause of crimes.[69] Under such conditions, young and single men are prone to risky behavior and are consequently responsible for the majority of crimes, especially violent ones.[70] There is an abundance of statistics that back this up. When compared with a woman, for instance, a man is twenty times more likely to be murdered by another man. This serves as a rough indication of the intensity of mate competition among men.[71] As such, David Barash believes that if we can get rid of male violence, we can eliminate nearly all violence.[72]

As early as the 1970s, biologist Richard Alexander argued that socially and legally imposed monogamy is a way of reducing sexual competition, solving conflicts, and lowering violence among men.[73] Data show that marriage can reduce the probability of male crime by 35 percent.[74] In China, as sex ratios (males to females) increased from 1.053 in 1988 to 1.095 in 2004, almost twice as many men became unmarried, and crime rates nearly doubled.[75] Across India, murder rates rose as sex ratios increased. In fact, they were two times higher in the male-biased district of Uttar Pradesh (with a sex ratio of 1.12) than in the slightly female-biased region of Kerala (with a sex ratio of 0.97).[76]

Science writer Matt Ridley argues that monogamy results from political compromise between rich and poor men in a society,[77] an assessment shared by Wright, who characterizes monogamy as "the grand, historic compromise . . . between more fortunate and less fortunate men."[78]

The question that remains is why the rich and powerful come to accept this major concession, giving up a significant reproductive advantage. When reproductive inequality becomes extreme in a polygynous society, the underprivileged and deprived men are inclined to rebel, as they have little to lose. It thus behooves the rich and the powerful men to compromise by curbing inequality as a way to control bottom-up violence and, more importantly, to prevent revolts that might overthrow them. Constraining polygyny, or better yet, instituting monogamy can pacify the potentially rebellious poor, who might otherwise have no wives, families, or progeny. Hence, compromising enables the privileged to keep the existing power structure intact, maintaining their other material advantages, which can still be converted into higher levels of reproductive success. This compromise exemplifies the logic of the Ultimatum Game, discussed in chapter 2. The best strategy for the proposer, Peter, is to make an offer that is generous enough that the recipient, Richard, agrees to take it. From his privileged position, Peter will lose more than Richard if Richard rejects the deal. Concession, rather than greed, allows Peter to maintain his *relative* edge over Richard. It is a smart move for the privileged to make.

However, monogamy per se does not wipe the slate clean of reproductive inequality. For instance, ancient Greek and Roman men kept concubines or female slaves for sex services in addition to their monogamous wives. In medieval Europe, rich laymen and clergymen often had many illegitimate children.[79] In medieval France, the right of a lord to deflower his serfs' daughters before their marriages gained a special designation: *droit du Seigneur*. In England, household servants also served the sexual demands of the masters, a practice that lasted well into the industrial age. Even Karl Marx, the father of Communism, fathered an illicit son with his maid. One does not need to be reminded of Arnold Schwarzenegger's sex scandal to know that the same pattern persists today. Whatever the era of human history, rich and powerful men across a wide range of cultures have almost always had greater access to sex, and thus an increased ability to pass on their genes to the next generation.

While the rich and powerful still hold an advantage regardless of their culture's marriage system, the social institution of monogamy does tone down the explicit reproductive inequality of polygyny. "Monogamy," in the words of Sigman, "then becomes redistributive to the extent it forces equality in the number of wives without solving the existing male inequalities in resources and characteristics."[80] Monogamy, helped by the church in preindustrial Europe, instituted as a social norm and sanctioned by law, greatly reduced the incentive for disadvantaged men to rebel in order to increase their reproductive opportunities.[81] As a compromise among males in sexual competition, monogamy is therefore interwoven with democracy, social stability, and egalitarianism—especially in reproductive equality.

Now we have finally established the driving force behind the overwhelming success of monogamy: fairness in reproductive opportunities for, literally, all men. It explains why the war to establish monogamy was waged more on the part of men than women in the course of Western civilization. This aspect of human history counters an overly simplistic and ultimately misguided view—that only women are victimized and deprived of their freedom to choose in polygynous societies. Had men—particularly poor and low-ranking men—not taken a strong interest in reproductive equality, it would have been difficult to imagine monogamy triumphing in the Western world. "Monogamy," in the words of Barash and Lipton, "is the great male equalizer, a triumph of domestic democracy."[82]

Although legalizing same-sex marriage has recently garnered a great deal of support, a substantial number of Americans still hold out. Some fear that legalizing same-sex marriage can be a slippery slope, leading to the legalization of polygyny in American society. The worry can be warranted only if we assume polygyny is simply a different lifestyle, comparable to same-sex marriage. In fact, many polygyny supporters do call the legalization of plural marriage "the next civil rights battle." This idea has attracted some high-profile proponents, such as Jacob Zuma, president of South Africa, and Carla Bruni-Sarkozy, the former first lady of France. Entertainment programs such as the HBO show *Big Love* and the TLC reality program *Sister Wives* have raised interest and increased support for legalizing polygyny.[83] Even some powerful legal organizations take a similar view. The ACLU defended polygyny in the name of intimacy liberty. "Talking to Utah's polygamists," asserts Steven Clarke, the ACLU's Salt Lake City legal director, "is like talking to gays and lesbians who really want the right to live their lives, and not live in fear because of whom they love."[84]

Is polygyny only a matter of lifestyle? Given full freedom of choice, who would you choose as your neighbor, a gay couple or a man with three wives? The difference lies in the fact that the former involves no fairness concern whereas the latter does. Do we want Kody Brown, his sister wives, and his growing number of children to become a widespread reality instead of a reality TV show? Unlike same-sex marriage, polygyny is not about alternative lifestyle, intimacy liberty, or religious creed; it is about reproductive fairness. Polygyny—a social institution that goes against the grain of our fairness instinct—is not an institution that *upholds* civil rights. On the contrary, it is an institution that *tramples* civil rights.

It is fascinating that during the heated legal battles over Mormon polygyny, from 1862 to 1890, all the prominent leaders who opposed it were male politicians. In the issue of women's suffrage, their silence was deafening. In contrast, women leaders in the suffrage movement at the time—Elizabeth Stanton, Susan B. Anthony, Lucy Stone—were silent on the issue of polygyny, rarely touching on it. So the anti-polygyny movement was almost exclusively a male endeavor. In the end, it was the overwhelming pressure from male political leaders and their supporters in American society that left the Mormons no choice but to renounce polygyny.

When the Mormon prophet Wilford Woodruff eventually announced the end of polygyny in 1890, he did not want his followers to be alarmed by his concession. He cautiously put down the following words in what is known as the Woodruff Manifesto: "Inasmuch as laws have been enacted by Congress forbidding plural marriages . . . I now publicly declare that my advice to the Latter-day Saints is to refrain from contracting any marriage forbidden by the law of the land."[85] He made no explicit admission of defeat.

Soon after the LDS Church gave up polygyny, the federal government returned confiscated church properties and pardoned convicted polygynists. Through six petitions, Utah ultimately won its statehood in 1896, after the provision that "polygamous or plural marriages are forever prohibited" was written into its state constitution.[86]

Even carefully couched in terms of God's will to preserve the LDS Church, the Woodruff Manifesto exudes a palpable sense of helplessness. While Christians welcomed and celebrated it as a landmark victory for monogamy, quite a few LDS insiders considered Woodruff a fallen prophet, kneeling to pressure from the US government.[87] Believing that the LDS Church had given up the cardinal principle of Mormonism, some radical fundamentalists broke away. To escape persecution, they went underground to continue their polygynous practice in remote areas of Mexico, Canada, and the United States.[88] At the time, few could foresee what was coming for the LDS Church in the post–plural marriage era.

Such a defeatist attitude among LDS members, if legitimate at the time, was soon proven entirely misplaced. The reality was that renouncing polygyny not only allowed the Mormons to gain Utah's statehood and a gradual acceptance of their religion by mainstream Americans; it also paved the way for

Mormonism's unexpected upsurge as a dominant player in the competitive spiritual market. In fact, in recent years, the LDS Church has been the fastest growing religion in America, an enviable position among all the major Christian denominations (see figure 1). Meanwhile, Mitt Romney, a Mormon, became a formidable challenger to Barack Obama in the 2012 presidential election. Though still underappreciated, the Woodruff Manifesto marked the most important turning point in the history of Mormonism—it has transformed an otherwise small, struggling spiritual community into a vibrant, fast-growing mainstream religion whose membership has been increasing across the world.

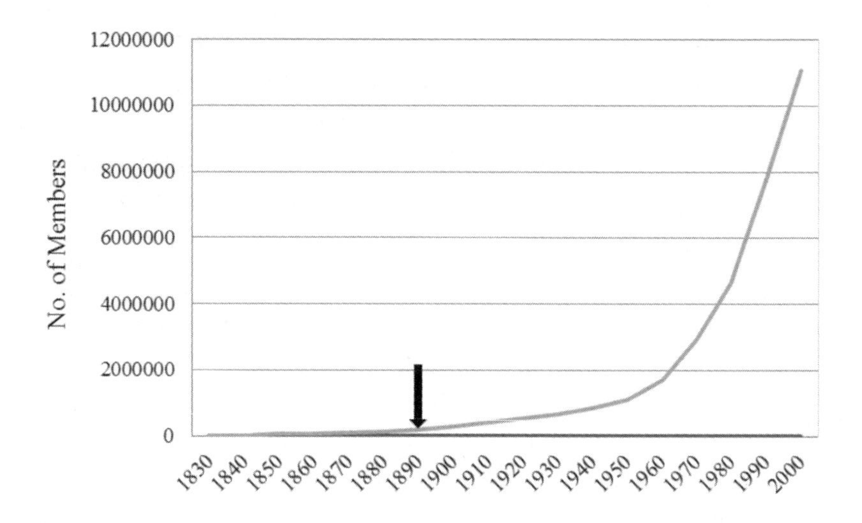

Figure 1. Membership growth of the LDS Church. (The arrow indicates the year when polygyny was renounced.)[89]

Since many branches of monogamous Mormonism fail to prosper as the LDS Church does, it is hard to say that monogamy is the only reason for the success of the LDS Church. But if it had not been for its rejection of polygyny, the LDS Church would never have thrived so dramatically. In sharp contrast, the FLDS Church has remained marginalized as a fringe sect in remote and isolated communities. Here again, the human desire for fairness, in this case for reproductive equality, has played a vital role in the prosperity of the LDS Church today. By extension, the phenomenal rise of Mormonism suggests that the success of Christianity may have also resulted from institutionalizing monogamy, which caters to the deep-seated human desire for equality, fairness, and justice.

As we learned in the previous chapter, Christianity played a major role in the long, hard fight for monogamy. It was a militant cultural force that stood in dogged opposition to the reproductive inequality that polygyny represented. However, there is still another major piece missing from our jigsaw puzzle. In the fight for reproductive fairness, why would its proponents need to appeal to the divine? Why does our natural evolutionary inclination need to be sanctioned by the blessing of God? These questions force us to face one of the most complex questions of our time: what is religion for? We will explore this question in depth in the final chapter.

Chapter 13

COMEUPPANCE AND THE HIGHER ROBIN HOOD

Exhorting the gods is a near-universal practice. Perhaps the presence of religion, laced with its notions of fairness and justice in nearly every human society, speaks to its deep evolutionary roots.[1] But why do we turn to divine beings to confer authority and legitimacy? "The human mind evolved to believe in the gods," comments biologist Edward O. Wilson. "It did not evolve to believe in biology."[2]

What purpose, then, does religion serve? Emil Durkheim, founder of sociology, saw religion as a way to promote solidarity within a community. Religion is "a unified system of beliefs and practices relative to sacred things," he wrote, "that unite into one single moral community . . . [for] all those who adhere to them."[3] In chapter 8 we learned that a major dilemma in our social lives is that mutual cooperation is often not stable whereas mutual defection is. This explains why friendships tend to be fragile—often difficult to forge, yet easy to ruin. Fortunately, humanity did not travel its evolutionary course as a single, isolated social group. Interactions among many different societies played an important role in shaping human behavior, relationships, and institutions.

As was illustrated in earlier chapters, had humans evolved in an isolated society, where relative payoff was all that mattered, our behavioral instincts would have more strongly favored competition with our peers over cooperation. Beyond spouses, children, and other close relatives, we'd have been more likely to treat distant relatives, neighbors, and others as competitors for scarce resources. This is the milieu in which the Spite Game can prevail and the strategy of mutual defection wins out.

But, thankfully, the character of that game changes when neighboring societies pose a threat to the very survival of our own society. In this situation, the collective strength of our society matters for all members because our individual fates are bundled together. The multiple small benefits that accrue from cooperative behavior within our group can add up to a tangible edge over our rivals. The advantages internal cooperation conferred may be the reason our ancestors earned their existence among rival tribes.[4] In short, competition against outsiders forces us to be more cooperative within our own group.

To maximize the benefits of cooperation, individuals in a given society must promote trust and stamp out spite—envy, hate, betrayal, backstabbing—among peers. More elegantly put, "united we stand; divided we fall." This task of unifying a society often falls onto the shoulders of religion. Not surprisingly, love and justice are universal values in all major religions in the world. Greater meaning or metaphysical overtones aside, the practicality inherent in these messages is evident: they promote group cohesion, harmony, and solidarity while circumventing tendencies toward infighting. Moral axioms like these inspire and encourage people to put away the kind of personal grudges epitomized by revenge and retaliation, which are forms of spite that can weaken a society. It is of little wonder that people are inclined to cling to the divine when their societies face external threats. In the 1970s and 1980s, for example, the Catholic Church in Poland and in Baltic nations became a rallying point in the struggle for reclaiming national sovereignty from Soviet control; divided rebel groups in Afghanistan united under the mantle of the mujahedeen (Muslim warriors who struggle in the name of Islam) to fight Soviet occupation; and revolutionary Iranians rode a wave of religious fervor that toppled the Shah's rule and broke Iran free from American and British influence.[5]

As we saw in previous chapters, the real threat to group cohesion comes from the parasitic free riders, who, by nibbling away at group strength and depressing the morale needed for cooperation, can make a society vulnerable to external hostilities. But free riding seems inevitable in any large society. And worse, as a society grows, it creates more opportunities for free riding, which can boomerang to dismantle internal cohesion, leaving the society prone to collapse.[6] Control of free riding, therefore, becomes vital for the survival of a society as it grows, and, for this reason, people increasingly turn to religion.[7] Indeed, a study of 186 societies reveals that the larger the society is, the more likely it is that deities are invoked to sanction the code of human behavior.[8]

Religious rituals serve the same end. "Whatever the origins of religiosity," points out psychologist Jonathan Haidt, "nearly all religions have culturally evolved complexes of practices, stories, and norms that work together to suppress the self and connect people to something beyond the self."[9] They consolidate shared norms and values essential for sustaining a sense of community. Research over recent decades has demonstrated their observable effects. Neuroscientists have discovered that reciting a Bible psalm can activate the prefrontal cortex, which may be linked to personal experiences of God.[10] By the same token, rituals such as Buddhist meditation can block information flow into the posterior superior parietal cortex and induce a surreal feeling of higher reality and unity while blurring the boundary between self and others.[11] And when Carmelite nuns have the mystical feeling of unconditional love and oneness with God, a higher level of activity can be observed in multiple brain regions.[12] Further screening shows that praying can activate the right caudate nucleus, which in turn stimulates the striatum, a brain region involved in reciprocity and social trust.[13] It should be no wonder, then, that when people pray, they tend to become more cooperative with other members in their communities.[14]

High levels of trust among peers bring real benefits to a society. By appealing to supernatural causes, prehistoric humans enhanced group solidarity, enabling cooperation in activities such as hunting, food sharing, defense, and warfare.[15] Similar cases can be found in historical records. For instance, trust that arose from shared religious beliefs apparently facilitated the expansion of trade networks among medieval Jewish Maghribi merchants and Muslims in Africa.[16]

Having a shared belief system can also facilitate rebuilding trust.[17] The practical utility of religion is reinforced in an analysis of two hundred communal groups in nineteenth-century America. In this study, anthropologists Richard Sosis and Candace Alcorta unveiled a major disparity in staying power between religious and secular communities: the former outlasted the latter by a ratio of five to one, indicative of religion's role in resolving internal strife in a society.[18]

A key to the survivability of religious groups seems to be their practice of rewarding goodness and punishing evil.[19] This, in turn, explains why most religious practices contain both carrots and sticks. On one hand, they advocate for love and forgiveness, which promote group cooperation and overcome internal strife that can weaken the society. On the other hand, and more importantly, they also enforce morality by meting out punishment—from social ostracism to physical chastisement—to those who violate rules.[20] Jews and Christians, for example, are assured that "sin will be punished,"[21] a familiar theme repeated numerous times in the Old and New Testaments to warn against cheating and defection.

Divine punishment can be perceived as real. In a survey of 186 traditional societies, every one attributed illness to supernatural beings.[22] The Old Testament is replete with similar stories. When the Pharaoh refused to let go of the Israelites, Moses, at the behest of God, rained down the famed seven plagues of Egypt, from locusts to fires, culminating in that fatal disease from which Israelites were protected by lamb blood painted over their doors. Thereafter, the Israelites were left free to wander into the Promised Land, and God spared Egypt for a time, as the Egyptians had learned their harsh lesson. Such a turn of events is not surprising, considering the dire fear of disease and illness in the past, when humans had little knowledge of how to fight disease beyond praying for divine deliverance—which gave birth to the Western salutation to one who sneezes, "God bless you!"

The fear that cheating and defection may bring divine wrath is a powerful deterrent to free riding. But to punish a sinner (in this case, a cheater or defector), one must first detect the sin. This partly explains why people are interested in each other's activities, both in real life and in fiction. "Stories are the medium for . . . monitoring," writes literary critic William Flesch, "and they are interesting because we are so tuned to monitoring one another, and how we monitor one another, and what stories we tell about one another."[23] Even so, it is humanly impossible to monitor everyone in a society *all the time*. Neighborhood watches—even by government use of extensive networks of community committees in the case of China—can only partially fulfill this function. But if an omnipotent and omniscient divine being can be tapped for the job, it's no longer a problem. This idea about the utility of a supernatural being is dubbed, quite seriously, the "God as policeman" hypothesis.[24] With God's constant presence, any secret action that may be harmful to society can in theory be detected and punished.[25]

The intimidating effects of an all-powerful, all-knowing, and always-judging deity can be dramatic. College students who wrote down the Ten Commandments before a test—a reminder of the presence of God—were less likely to cheat.[26] Young children were better at following rules when told a supernatural being was watching over them.[27] Michel Foucault, a popular French social theorist, argued that it is the internalized feeling of potentially being watched that makes one less likely to commit a crime.[28] Apparently, he was right—hanging up a pair of graphic eyes makes people more likely to pay for coffee in the workplace, as shown in a prankish experiment conducted at a British university (in many universities, coffee is supplied based on an honor system by which people voluntarily pay to an "honesty box" as they drink).[29] With an all-knowing divine being, anonymity ceases to exist and secrecy is banished. Any action may affect one's reputation, as if one is under constant surveillance by a video camera or the police. But more than that, an omniscient god knows you inside out; he reads your deep desires, intentions, and thoughts. So you'd better behave yourself with absolute sincerity.

Besides moralizing and punishing gods, religion offers another powerful incentive—the afterlife, which, in a sense, becomes real when there are enough people to believe it.[30] The afterlife is a feature of all major world religions, and it is also broadly conceptualized among tribal peoples. Some archeologists consider ceremonial burials evidence for this belief. In that sense, Cro-Magnons living fifty thousand years ago had already conceived of the idea.[31] Apparently, notions of the afterlife in Judeo-Christian tradition or of reincarnation in Hinduism and Buddhism serve practical goals: they encourage people to behave well in this life—to be honest, work hard, take responsibility, abide by law and order—in exchange for the promise of a glorious life in the next.[32]

Bolstered in part by the fear of eternal judgment, religion and its notions of justice are useful in a very real way. But there is one fundamental problem:

divine justice is a comeuppance we *believe in*. Reality, as is often the case, may not go along. Wrong, treacherous, and evil acts may just as well go unpunished or even be rewarded while natural or human-made disasters can hurt good and devout people. Gods—omnipresent, omnipotent, all-knowing, and all-fair, as they are commonly believed to be—can apparently fail in their presumed earthly duty. This cognitive dissonance has been debated for centuries, if not longer, and it has sparked a cornucopia of philosophical answers. Among Christians, it has often been reconciled through reinterpretation: humans can be wrong in understanding God's will; God must have other, grander plans beyond worldly comprehension. Alternatively, God may be using disasters to test the strength of followers' faith. Or such disasters can be interpreted as divine wrath intended to punish people for their sins.

When the divine will is perceived as running contrary to the service of justice, many are vulnerable to experiencing a substantial loss of faith. A few decades ago, a survey found that 63 percent of Americans became angry at God after suffering trauma, loss, or other adversity. Some even went further than blaming God for causing their misfortune. Among college students who had experienced a major negative event in life, 21 percent lowered their confidence in God, and 9 percent rejected religion altogether, becoming atheists.[33] We can see that the relationship between the human and the divine is guided and governed by the fairness instinct.

Why are most religions rich in taboos and moral judgments regarding personal conduct, particularly sexual practices? What does regulating personal habits have to do with promoting group cohesion? To answer these questions, consider this: gods are supernatural, to state the obvious. Part of their mystique is that they are beyond our normal sensory experience. We can only understand and describe them with our limited faculties. For Daoists, for instance, gods can be a bevy of human specialists with superpowers; for Buddhists, Buddha has three heads, six arms, and numerous morphs; for monotheistic Judeo-Christians, God has been perceived in a bewildering variety of ways, from an old, wise, white-beard man to a bodiless yet omnipresent force. Unfortunately, defining the nature of the divine is often a task laced with potential for explosive conflict. For example, the centuries of theological debate throughout the classical world on the exact relationship between God and Christ led to wars and rebellions, eventually paving the way for the split between Orthodox and Catholic Christianity in the eleventh century.

The magnitude of these religious conflicts is a testament to the difficulty in deciding which interpretation of the divine is "correct." While this is an important theological question in its own right, behind it is a more practical concern with looming implications for human societies—who exactly is qualified to interpret and communicate divine wills? Political leaders have often tried to assume this mantle. Throughout human history, kings have invoked divine sanction, emperors have declared themselves gods or descendants of gods, and caliphates have asserted the right to joint spiritual leadership by virtue of religious doctrine. Regardless of their personal religious or philosophical beliefs, leaders are, more often than not, painfully human; they have their own vested interests and can become corrupt. Hence few wonder why political leaders have so frequently been regarded with suspicion and are ranked so low in trustworthiness in recent surveys of public opinion.[34]

So how can one win trust from others? Diverse approaches aside for garnering a good reputation, there seems only one surefire way to show selflessness—to give up personal interest and desires for the benefit of the community. As such, priests, regardless of their intensions, often deliberately handicap themselves by living in harsh conditions, denying human comfort and biological desires. How, then, can such rituals serve as an indicator of trust? An insight to this question rests in what seems an irrelevant observation: the peacock's tail.

Evolutionary biologists since Darwin have been baffled by the same question: why does the peacock grow a gaudy yet utterly useless tail; how does this help the peacock to thrive? Few convincing explanations emerged until 1975, when Israeli biologist Amotz Zahavi proposed an elegant model known as the handicap hypothesis.[35] The idea is simple. Females are choosy, and males are selected to win over finicky females. Since good genes are, by definition, good for the survival and reproduction of offspring, females should be attracted to males who carry good genes. But genes are hidden. How can a peacock persuade a peahen that he has good genes? Unfortunately, he has no alternative but to let his tail—an onerous and debilitating burden—speak for the quality of his genes. In other words, he has to put his money where his mouth is: the larger his tail, the better he can survive, the better genes he carries. Since an extraordinary handicap—the splendid tail of a peacock; the massive antlers of a bull moose; the loud, deep, and persistent croak of a male frog—often requires high energy inputs while inviting so much more risk, it is beyond the ability of the less fit of the species to fake it. The logic is comparable to playing chess with a handicap. The greater the handicap one can overcome in order to win, the greater the display of skill. Though costly, a handicap is by all means honest.

The handicap hypothesis is supported by evidence in a wide variety of species, from tiny insects to large mammals, and in many contexts beyond sexual advertisement. The handicap hypothesis has a corollary in communication known as the costly signaling theory. There is no shortage of examples in humans in this regard. For instance, why do men on average have a larger body size than women, who, as it so happens, tend to prefer tall, muscular men? Perhaps, goes one plausible answer, strong men were able to provide better protection in our evolutionary past. But big and robust men also require more food. Satisfying this requirement is difficult when food is scarce, a perennial condition under which humans evolved. Hence, body size and musculature might have also served as physical handicaps. So, too, bravery in combat or during hunting expeditions may have been a behavioral handicap. Among fun-seeking college students, gallant chicanery such as jumping into cold water or downing numerous bottles of beer, performed in the presence of females, may have the same functionality. Some native peoples go further afield in displaying handicaps during courtship. The Dogon of Mali, for instance, put bulky masks on their heads to sport their physiques during festivities. Because masks can be very large and heavy, only the strongest men can adeptly handle the largest masks during dancing rituals.

To gain others' trust, we often do something similar to donning the Dogon masks. Notice that temples, mosques, and churches are often the most impressive structures in a community. Poor villages and shanty towns still manage to pool scarce resources to sponsor such buildings, even though they're of little practical use. When resources are plentiful, communities construct religious shrines that are bigger and better and more meticulously detailed, through which the strength and piety of a community is conveyed. Like the out-of-control competitive process that led to the male peacock's burdensome tail, the drive to display piety through extravagant architecture can sometimes run amok. Notre Dame in Paris, for instance, took about eight decades to complete, and St. Peter's Basilica in Rome took 120 years to build. These were ultimate handicaps, exhibiting ardent human devotion to the divine. So, too, are grand religious structures and relics in Athens, Rome, Istanbul, Xian, Angkor, Mecca, Varanasi, Jerusalem, Guatemala, and many other ancient sites.

Trustworthiness is advertised in a similar manner. Like the peacock's tail and the building of audacious religious monuments, trust is erected through self-sacrifice, in many cases by eschewing mundane comforts and desires. This explains why many religions are rich in taboos that specify prohibited behavior, particularly in regard to food and sex. Such rules usually confer no clear biological advantage in a conventional sense. In the case of Buddhism and the Greek Orthodox Church, many shrines and monasteries are located in remote mountains. Apart from the arduous and dangerous construction work, life there is anything but easy. But this is exactly the point. By choosing to live and work there, monks exhibit their purity, devotion, and determination.

By the same token, religious followers are required to devote a significant amount of time and energy, with considerable material costs, to demonstrating loyalty and commitment.[36] As a case in point, the Amish handicap themselves by rejecting modern technologies to promote cooperation and strengthen the fabric of the community. Muslims fast during the entire month of Ramadan, to a degree that their fasting interferes with their performance in major sports events such as the Olympic Games and the FIFA World Cup. Likewise, they spend a sizable portion of their income and time traveling to Mecca to fulfill their religious duties. Christians are required to shun the seven deadly sins—envy, lust, pride, greed, gluttony, anger, and laziness—which are a litany of basic human desires and emotions. This list of examples is not exhaustive, but in this shared notion of sacrifice we find a unity among the cacophony of religious rituals, taboos, obligations, and other practices, such as hard work, tithing, charity, asceticism, fasting, and celibacy. They are self- or community-imposed handicaps that evolved from, in no small part, the drive to win the trust of peers, a reason why early Christians took Jesus' injunction to "give up all you have and come and follow me" quite seriously.[37]

Now we know why religious rituals are often lengthy, meticulous, onerous, and, most important of all, hard to fake: they are ideally suited for symbolizing and conveying honesty and trust. This explains an enigma: the more stringent a religion's ritual requirements, the more committed its adherents tend to be. Indeed, in terms of church or synagogue attendance and donations, groups with strict rules and practices such as the Mormons and the Orthodox Jews outdo the more liberal Methodists and Reform Jews, respectively.[38] Accordingly, numerous religious groups insist that new members go through onerous rituals, many clearly detrimental to their well-being. In the same vein, religious codes often detail practices, taboos, rules, and mandates about food, charity, work, dress, sex, ritual, and other aspects of daily life and activities. An exhaustive survey on the ritual practices in nineteenth century America shows that religious communities imposed more than twice as many costly requirements on their members (such as food taboos, fasts, marriage, sex, and limits on material possessions and communication with outsiders) as their secular counterparts.[39] Not surprisingly, members in religious groups are more likely to provide help to one another when compared with those in secular groups such as parent-teacher associations or bowling leagues.[40]

Similar strategies are used for the same purpose by some secular organizations, including college fraternities and military boot camps: neophytes are tested with tattooing, beating, genital mutilation, being forced to eat toxic substances, and deprivation of food, water, or social contact.[41] Paradoxically, these abuses are supposed to help build a close-knit society.

As highlighted in the last two chapters, reproduction, in particular, weighs heavily on our sense of fairness. For this reason, sexual morality is among the most visible issues in virtually all religious communities. It can win followers' trust, if done properly, or drive them away. As such, sexual acts and relationships that are deemed improper—adultery, polygyny, promiscuity, prostitution, homosexuality—are made taboo. Among the best ways to demonstrate trustworthiness is simply to willingly forgo sex. This provides an explanation for why monks and nuns depart from villages, towns, and cities and spend their time in desert and mountain monasteries cut off from the world around them, while Catholic priests and Buddhist monks give up marriage to serve a higher power. In some situations, the entire religious community practices celibacy. Shakers, for instance, refrain from sex and, as a result, depend on converts to avoid extinction.[42]

Here, much like the peacock uses the onerous tail to speak to the quality of his genes, celibacy, regardless of its practitioners' unspoken intentions, speaks to an honest commitment to a greater good. Those who pass this test are seen as being above the world, more trustworthy and altruistic, and accordingly fit to lead. Conversely, improper sexual relationships for religious leaders—and for politicians—serve the exact opposite function. They represent a breach of trust and smear the image of the church that represents the professed faith of the disgraced leader. Sexual abuse scandals in recent decades have become a nightmare for the Catholic Church in Great Britain and America.

So far, so good for trust. But there is bad news. Sometimes, even though we have gone to great lengths, handicapping ourselves to convince people that we are honest, loyal, and altruistic, we may still fail to dispel all clouds of suspicion and doubt. One way to overcome this difficulty is to "outsource" to divine beings, who are entirely detached from worldly concerns and can therefore be absolutely fair and ultimately just. Such "external help" can aid in reducing the level of suspicion and doubt among members in a community.

Resorting to divine power, in fact, is a universal phenomenon. Ancient despotic Chinese emperors claimed themselves to be the sons of heaven to legitimize their rule. Aztec kings presided over the ritual of sacrificing human lives to please the sun god who kept the world moving. Judeo-Christian and Muslim societies have developed laws under the sacred name of God to beef up their authority and legitimacy. Even secular icons, institutions, and events are sometimes consecrated with a tinge of divinity. Confucianism, for example, has been explicitly nonreligious. Confucius never claimed himself to be divinely ordained. His teachings have been followed as a philosophical rather than a religious system. Yet many Chinese traditionally worshipped statues of

Confucius as if he were some sort of god. The cases can go on and on with the same leitmotif: divine power is invoked to install a sense of infallibility and credibility. Infusing the institutions of power with divine sanction, therefore, lends an uncontestable legitimacy and authority, enabling those in power to preserve order. Consequently, leaders in inegalitarian societies, compared with their egalitarian counterparts, are more likely to invoke gods to justify social control.[43]

This evidence shows that there is a strong force behind the evolution of human religiosity: the lingering mistrust regarding human motivation. It leads to the common belief that only the divine can be absolutely fair and just. This explains, for instance, why Islam allows only Allah to claim infinite justice, and why we so often invoke God when dealing with serious and consequential issues. Not surprisingly, modern national anthems are laden with appeals to divine blessings. In fact, among 195 national anthems in the world today, fifteen have the word "God" in their titles, from "God Bless the Sultan" (Brunei) to "God Save the Queen" (the United Kingdom); many more contain the word explicitly or implicitly in their lyrics.[44]

In the United States, the word "God" crops up in a bewildering assortment of political, economic, military, and legal speeches and ceremonies. In virtually all serious places and most solemn times, vast numbers of Americans pray to God for his mercy, power, love, blessing, fairness, and justice. The Pledge of Allegiance requires citizens' unreserved dedication to the nation "under God," a phrase that was missing in the original version but was deliberately added during the Cold War in response to the supposed "godlessness" of the Communist threat. People swear to tell the truth in courts, traditionally done with one hand on the Bible. The government inscribes "In God We Trust" on coins and bills. In presidential inauguration ceremonies, the phrase—"So help me God"—arguably a slip of George Washington, has been followed verbatim as an essential element of the oath of office ever since. US foreign policies, especially in regard to warfare, are sometimes couched in the rhetoric of religious duties to defending good against evil to drum up popular support.[45]

It seems that the popularity of divine beings may mirror the status of a nation in social, political, economic, and legal equality. Laws and legal codes, in particular, are an interesting arena for expressions of the divine. Gods are often thought of as lawgivers. In Judeo-Christian and Islamic traditions, laws were often written and enforced under the name of God. The controversy between natural law, which is given by God or inherited from nature, and positive law, which is imposed by humans, has divided philosophers for centuries, if not more, in the West.

In the United States, modern legal codes come from a historical legacy of ecclesiastical and canon law. Resorting to the divine has been a common ritual in judicial and legal practices, despite their secularity. The American Supreme Court is adorned with a statue of Moses holding the Ten Commandments, and the legislature opens with a prayer for guidance. In the historical dispute about the legitimacy of Mormon polygyny, God was consulted by both sides. "The congressional debates on polygamy in the nineteenth century revealed a number of reasons why the wider public opposed it," writes Philip Kilbridge. "For one thing, polygamy was seen as a violation of natural law, that is, God-given monogamy." As such, polygyny was "unethical and akin to slavery," for which it had to be eliminated.[46]

Unfortunately, despite the general expectation that divine beings are absolutely fair and just, the real world rarely yields such idyllic solutions. So in the place of God sitting in perfect judgment, we settle for divine inspiration and guidance, hoping they can help us avoid human errors and biases.

Divine sanction is a double-edged sword. Just as rulers appeal to divine authority for exercising power on earth, the poor and disenfranchised appeal to divine justice to protest against its excesses. These two faces of the same appeal lead to dynamic relationships between those who rule and those who are ruled over. For the higher-ups in a hierarchical society, claiming divine justification for their power allows them to consolidate their privileged positions in the regime because the divine will cannot readily be challenged. For the lower-downs, meanwhile, divine justice can be a rallying call for demanding a compromise from higher-ups toward equality. Unsurprisingly, among the strongest appeals of the Judeo-Christian and Islamic religions is their claim that people are equal under God. Divinely ordained equality has proven a major attraction for those who sit on lower rungs of society, and, as such, the poor and the powerless are often more pious and devoted to God than the rich and the powerful.

As we saw in Chapter 2, this bottom-up pressure for fairness in human societies can become so strong that pleading to gods—as did many ruthless despots in the Chinese, Aztec, and Incan empires, as well as those throughout Europe and the Middle East—may not be adequate for leaders to legitimize their rule. Christians, for instance, place their duty to obey God above their duty to obey a human ruler.[47] This strong religious devotion, which can inspire subjects to defy their rulers regardless of the consequences, often takes higher-ups by surprise.

Early Christians in the Roman world took literally Christ's admonishment to "Render unto Caesar the things which are Caesar's, and unto God the things that are God's" (Mark 12:17). In that vein, they were encouraged to be good citizens who paid taxes and abided by the laws of the Roman state, so long as Roman authorities did not force them to contradict the tenets of their faith. When the Roman Emperor Diocletian ordered that all subjects must honor his and the state's divinity with a sacrifice, Christians refused. Their defiance toward acknowledging Roman divinity marked a new wave of persecution among Christians. Even so, the realities of the injustices of the late Roman world sparked in many an intense interest in Christianity, which promised an afterlife that would reward those who led good lives in the middle of such depravations. This growing influence amid political persecution ultimately bore astonishing results. Less than half a century after Diocletian died in 311, those who had formerly been persecuted for ignoring the emperor's divinity ended up triumphant once the emperor Constantine converted to Christianity himself and set the vast territory on its course to becoming a Christian empire. And the rest is history.

Diocletian already noticed that religion can work both ways for the state. While it gave new legitimacy to his rule in a brutally unfair and chaotic world, religion equally fueled resistance among those who declared their allegiance to a higher power against whom no man could contend. This legacy flourished centuries later in Western Europe when claims of divinely granted or naturally existing equality and liberty became fundamental drives toward civil society and democracy. Just as Diocletian and countless other leaders assumed the mantle of divine sanction in order to rule, those over whom they ruled asserted in the name of God their right to protest or revolt against actions or rulers they saw as unjust. This antagonism has been on the rise since the Protestant Reformation, when debates about the nature of the divine led to nearly a century of religious wars that ended with the Peace of Westphalia in 1648. But the real fruition of the divinely inspired right to rise up would come with the Enlightenment that followed.[48]

On July 4, 1776, the Atlantic colonies of the British Empire united in revolt against the British Crown by sending off their Declaration of Independence:

> When in the Course of human events, it becomes necessary for one people to dissolve the political bands which have connected them with another, and to assume among the powers of the earth, the separate and equal station to which the Laws of Nature and of Nature's God entitle them, a decent respect to the opinions of mankind requires that they should declare the causes which impel them to the separation.
>
> We hold these truths to be self-evident, that all men are created equal, that they are endowed by their Creator with certain unalienable Rights, that among these are Life, Liberty and the pursuit of Happiness. —That to secure these rights, Governments are instituted among Men, deriving their just powers from the consent of the governed, —That whenever any Form of Government becomes destructive of these ends, it is the Right of the People to alter or to abolish it.

In one rather short document, Thomas Jefferson, who penned the declaration, brought into brilliant synthesis the core arguments of the philosophers and theologians of Enlightenment on the nature of rights, authority, and the individual. Now the religious tenets regarding the inherent equality of all people, which made Judaism, Christianity, and Islam so appealing, saw political actualization in the New World: not only are we born equal, we are endowed by our "Creator with certain unalienable Rights," and only by protecting these can a government rule over us. These concepts would not only come to define the American experience; they would also come to represent an invocation of the flip side of divine sanction, which, along with the power to enshrine the authority of a government, also has the power to take it away. This concept would, for the first time, give those out of power the ability to force a compromise and sue for fairness and justice.

Since the United States enshrined these two faces of divine power, "Nature's God" or "The Creator" in its founding document, it has been invoked to justify human rights. Though the drafters of the Constitution left out direct references to God, divine power and divine justice have been used countless times to justify American-style conceptions of democracy and human rights. It was in the name of divine justice that slaves gained their freedom; it was in the name of divine justice that women fought for and won their suffrage; it was in the name of divine justice that Martin Luther King Jr., himself a Baptist minister, led African Americans in their fight for civil rights; it was in the name of divine justice that Cesar Chavez led the way for the rights of Latin Americans.[49] In each of these movements, those who were kept out of power and marginalized in society were able to use the Enlightenment concepts of fairness and equality to force a compromise in the social and political hierarchy of the nation.

We should be clear that divine justice is metaphorical and metaphysical. As it can be claimed by the privileged—including bloodthirsty despots—for maintaining the status quo as well as by the underprivileged for equal rights and fair distribution of resources, divine justice is by no means a tangible guarantor of fairness. Although nearly all major civil rights movements in America have appealed to God, the ultimate guardian of fairness lies in our secular legal institutions. Here and in many other cases, divine justice serves as one psychological motivator that fuels and unites bottom-up struggle for equality and fairness in society. Make no mistake, the mantle of divine justice, as powerful as it can be, is not indispensable in bottom-up struggles for equality and fairness. In the French Revolution of 1789 and many Communist movements in the twentieth century, for instance, revolutionaries toppled ruling classes, along with their clerical allies. In these cases, the role of the divine was inconspicuous; perhaps more importantly, the religious systems supporting the old order were thrown out altogether. What rallied and fed these movements was the explicit expression of equality and justice—the very core of the human desire for fairness decoupled from the sanction of the divine.

During the civil rights movement in the United States, churches were often used as organizational bases and gathering venues. "The framework of the movement was enlightened," writes religious historian Garry Wills. "The power of it was Evangelical."[50] This case, along with many others, illustrates how human pursuit of fairness in worldly affairs can be affected by religious institutions. But this only tells half of the story, as human desire for fairness in the name of divine justice can also affect religious institutions. The latter process is best showcased by the Protestant Reformation, among the most important milestones in the history of Christianity.

On October 31, 1517, a little-known priest and professor of theology named Martin Luther hammered a list of ninety-five theses on the efficacy of indulgences to the door of the Castle Church in Wittenberg. A devout Catholic, Luther was concerned with growing abuses within the church that were undermining its sanctity and eroding the trust his German compatriots placed in the church and in the pope. People were being fleeced by the very shepherds who had vowed to guard their flock. "No central government protected the German people," writes Richard Marius, due to the inadequacies of the patchwork of states that made up the Holy Roman Empire, "and complaints against papal exactions rumbled out of Germany throughout the fifteenth century to Luther's time. Germans involved in legal problems concerning clerics or church lands found them remanded to Rome and faced an arduous journey across the Alps to have the case heard before clerical judges whose fairness to foreigners was doubted."[51]

Across Europe there were problems as well. Particularly unjust at the time were tax exemptions for clergymen and the promotions of "utterly stupid and incompetent persons" by bishops and abbots,[52] the guardians of the church, who were entrusted with the responsibility of modeling exemplary Christian lifestyles. In addition to selling holy offices, priests often demanded payment for performing masses, and officers sent by the pope extracted every last ounce of treasure from the locals, especially in the form of selling remissions from sins and assuring a place for believers in the afterlife. This last point particularly troubled Luther. Not only did it strike him as being morally and theologically questionable, but it also hit at the very heart of his beliefs in Christian fairness and the duties imposed on the pope, the man closest to the divine on earth. In a basic sense, a believer should obtain the reward of salvation for obeying their kings and the church as well as practicing all the "handicaps" their faith imposed on them. But when anyone could simply buy their way out of punishment—an afterlife in hell—it no longer mattered how they had lived their lives. As we saw in the capuchin monkey, creating an unequal reward system is risky business that can spark very primal anger because it violates the evolved sense of fairness. The Catholic Church would soon find this out in a very painful way.

As devout a scholar as he was a believer, Luther had undertaken extensive study to understand precisely what the scriptures—in his mind, God's direct revelation and the central guide to living a Christian life—had to say on the issues. What Luther found outraged him. He could find no religious sanction in the Old Testament or the New Testament for what the clergy was doing. In the name of tradition, many of the most powerful in the church had lived lavish lifestyles, disobeying prohibitions on marriage and sexual propriety, and gathering great estates and fortunes. They assumed their authority by claiming themselves the only group who truly knew what God wanted on earth. This only made Luther's outrage stronger. True, the behavior of priests and church officials was supposed to be above questioning, but not by virtue of their position; they were supposed to exemplify the highest standards of Christian behavior and live a lifestyle built on the sacrifice of personal desires—sex, wealth, power, food—for the good of their congregations. Luther understood that for the divine sanction of the Catholic Church to have power, it was not enough for it to rest on tradition and authority alone—the clergy had to live according to the dictates of scripture, and, moreover, the people had to actually believe they were doing so in order to trust in the divine sanction of the church.

The impetus for Luther's theses came from the behavior of the church under the newly invested pope, Leo X, son of the wealthy Medici family. Leo was seeking to raise money for the building of St. Peter's Basilica in Rome. Previous pontiffs had left the Vatican with some political strength and territory, but little in the coffers of the church. Leo, set on building a vision of heaven on earth, was determined to make Rome the center of the western world once again. To that end, he offered a series of special indulgences, church-sanctioned pardons. As Luther saw it, believers were being enticed by dishonest clergy to buy salvation for themselves and their dead loved ones under the guise of giving alms to the church. As a result of the campaign to raise funds for the building of St. Peter's, money poured into Rome, while the poor faithful were saddled with yet another burden in their quest for salvation. Luther decided to speak out on the abuses, confident that the holy father could not be party to the abuses of the greedy rogues who claimed to act in his name. He posted ninety-five points detailing the theological and doctrinal issues with the sale of these indulgences, as well as the toll they took on the believers who bought them. Believing that Pope Leo would right the wrongs if only he knew what was being done in his name, Luther dedicated his explication of the Ninety-Five Theses to Leo himself.

Given his position and education, as well as his belief in the truth of Christian faith, Luther had entered the fray and denounced these abuses in such a way that would remind the church of its true path and calling. For the average peasant though, these abuses were simply part of daily life and could not easily be contested. The Catholic Bible, the source of the church's authority and teachings, was written entirely in Latin. By the sixteenth century, this was an ancient tongue from a long-forgotten age, accessible to only a few. The average churchgoers still respected and venerated Christ, the Holy Mother, and the saints of the church, but many had grown to distrust the clergy.

By Luther's time, corruption all the way up the hierarchy of the church had been rampant. Pope Alexander VI, for instance, "managed to bribe his way to victory in the election to the papacy in 1492 despite the awkwardness of having several mistresses and at least seven known illegitimate children."[53] Rome itself appeared to be anything but a vision of the City of God:

> The city swarmed with prostitutes, some living in elegant palaces, frequented by members of the high clergy and treated as grandes dames. They came from everywhere in western Europe. Homosexuality among the clergy

> was common, acknowledged by many Italians, its practice by clergy high and low later condemned by Pope Leo X. Pope Julius II was said to suffer from syphilis, the new disease from the New World, and he was accused by some close to him of homosexuality. The streets were made dangerous by beggars, many of them vagabond monks crowding into the city to live off the tourist traffic.[54]

For centuries, the church had claimed that its teachings were infallible, but the unceasing corruption, unfair extraction of resources by the clergy, and the increasing involvement of the church in affairs of state made many begin to reconsider the status of the Roman Catholic Church as the one true universal church.[55] Thinkers of the high Middle Ages like William of Ockham had seen these human failings even before Leo's campaign to sell indulgences had brought the issue to a head. They began to question the shared belief in the irrefutable authority of the Catholic Church, which was predicated on nothing more than claims of ancient authority and tradition: "Faith was preserved not in a hierarchical institution but in a community of the faithful dwelling in the institution but sometimes not occupying its high and visible offices."[56] Such an idea—that faith thrived not in institutions that claimed divine right but in the shared belief in divine power—would eventually propel Luther into questioning not only the behaviors of men who shielded their bad behavior by using their power within the church, but also the legitimacy of the pope and the church itself.

Printed and reprinted after being translated into German, which could be understood by the layperson, Luther's writings went viral. "The reason for their overnight popularity was not so much theological as economic and national," Marius points out.[57] "His message and appeal aimed at a democratic understanding of the church in which all Christians stood on equal ground as priests."[58] The church, alarmed by this rise in Luther's popularity and the growing support for his teachings after their initial publication, asked Luther to recant. As a result of his refusal, he was excommunicated in 1521. Freed from trying to urge the church to reform, Luther was left with the series of vexing questions we encountered at the beginning of our discussion: What is the nature of divine authority? Who can possess it? And who is qualified to speak to it, and for it? Luther again returned to where his critique of the church began—the scriptures.

Hiding from church and civil authorities in Wartburg Castle in 1521 and 1522, Luther compiled a German version of the New Testament. This new Bible was translated directly from the original Greek, not the official Latin translation done some twelve hundred years earlier by St. Jerome. Written in the vernacular, it allowed ordinary Germans to find the faith directly, without relying on any priest or pontiff who claimed to have divine sanction to interpret the teachings of God. Luther's doctrine of *Sola Scriptura* (meaning "by scripture alone") represented a remarkable democratization of faith. Returning authority to the scriptures would become the defining feature of Protestantism. In this new conception of authority and personalized faith, "hierarchy would disappear."[59] Obviously, "his emphasis on making the Word of God accessible to all deprived the elite class (the priests and some rulers) of control over words as well as the Word."[60]

In challenging the doctrine of priesthood, Luther struck a chord with ordinary Germans. For him, the priest "only affirmed what God had already done."[61] He declared that "God, not the party and not the church, is sovereign in history."[62] Ultimately, Luther and is followers had come to believe that "the true church was not an institution, but a communion embracing all who confessed Christ; it was not centered in Rome but rather present wherever faith lives in the human heart."[63] In breaking with Rome and seeing themselves "no longer . . . cravenly subservient . . . to the clergy,"[64] these Protestants (literally, protesters) ignited a religious revolution—the Reformation—centered on what theologian Alister McGrath calls a "dangerous idea." It would give all Christians the right to interpret divine will and authority for themselves, resulting in "the development of Protestantism as a major religious force in the world."[65]

The Protestant Reformation would have long-range consequences far beyond what anyone had imagined, but it had its roots in fairness. Protestantism began its revolt when many began to lose trust in the clergy, their spiritual leaders, who had fallen prey to base human temptations at the expense of the community of believers. This loss of trust led to a loss of faith in the spiritual authority that the church and the pope claimed as the source of their power, pushing Protestant reformers, like Luther, to find a "truer" source of authority to determine the nature of the divine, and who was fit to interpret it and claim sanction from it. As a result, this loss of faith fractured communities and set the stage for the humbling of the Catholic Church, which had held Europe together for over a millennium.

There are countless other examples of debates about what is just and what is fair—far too many to detail in a single book, let alone a single chapter. However, the breadth and diversity of examples that stretch from prehistory to modern-day social policy, embodied in figures such as Diocletian, Constantine, Luther, and Jefferson, illustrate the deep human connection between fairness and religious faith. Before ending the chapter, I present a quote from psychologist Matt Rossano that recounts the origin, evolution, power, and function of religion far more elegantly than I can:

> Religion is paradoxical. It emerged out of intergroup competition, yet it has a remarkable power to unite. It sets people apart from one another, yet can also draw vastly different people together. Long ago we envisioned relationships with spiritual beings. In one sense these relationships were very mundane—they served practical human needs. The practical side of these relationships is also the source of their power to divide us, to pit one group against another. The divine side is in their capacity to quietly but persistently call us to transcend ourselves and find common ground for trust.[66]

But behind the paradox, we see why religion can be popular. Here, the most appealing part of religion is its *promise* to eradicate human flaws, biases, prejudices, hierarchies, and all other inequalities and injustice. Once again, fairness proves to be an essential ingredient in our religious lives. God is the ultimate Robin Hood; God's will embodies, distills, and transcends people's will. Our conception of heavenly comeuppance, therefore, is an outgrowth of our fairness instinct. In this sense, Voltaire's dictum over two centuries ago—"if God did not exist, it would be necessary to invent him"[67]—still rings true today.

Epilogue

THE FUTURE OF FAIRNESS

A rich village landlord prays to God every day, asking to be granted a wish.

God is moved by his piety and asks, "What do you want?"

"I wish the world to be mine, and mine alone. I don't want to have to share it with anybody else. I don't want to deal with people; I am tired of them."

"Then, it's yours."

Then, the man has a second thought. "But, I need a wife, too."

"No problem. Here she is. What else do you need?"

"I think . . . I have everything now." No sooner has he completed the sentence than he starts having regrets. "I guess I like fresh whole-grain bread baked by the guy at the bakery down the lane. So, if possible, I would like to have that fellow in the world as well."

"Certainly, your wish is granted. Is that all?"

Again, the man hesitates. "Making bread requires flour. So I need a farmer to grow wheat . . . and probably a miller to make the flour."

"Ah-ha!" God begins to laugh. "Then you'll also need a blacksmith to make tools for the farmer and the miller. And then a tailor to make clothes for the blacksmith. And then a weaver to make cloth for the tailor. And so on, plus their wives, children, friends, and relatives. Now, you are back to the same world with the same people. Are you sure you want me to do these all over again for you?"

Embarrassed, the man realizes that he can't live alone.

This allegory feels like a replay of Adam Smith's concept—that the world functions through economic relationships among people. But there is more here, because families, friends, relatives, and neighbors are all factored into the equation. Economic networks aside, evolution has endowed us with a strong affinity for social and emotional ties, so strong that few can live without them. Through the eyes of a shipwreck survivor, Daniel Defoe imagined what it would be like for someone forced to live alone. Readers (and viewers of the many movies and TV series based on the same story) breathe a giant sigh when Robinson Crusoe returns to our society—the same society described in the allegory above. The long, lonely suffering of the desolate, deprived man is finally over.

As recounted in chapter 2, humans, like many social mammals—such as primates, canids, and cetaceans—have evolved to be social. This is not simply a preference for the company of others, but a bond so vital to our physical and mental well-being that we can no longer be detached. Even in Western prisons today, solitary confinement—complete isolation from peers—is the worst punishment a prisoner can get. Being part of a social fabric is critical to our physical and psychological well-being. Despite conflicts of personal interest in our lives, fraught with human shortcoming and frustrations, we manage to subdue the conundrum of social living thanks to a fundamental instinct—the fairness instinct—that is part of our evolutionary endowment. From this instinct emerges a simple and elegant rule—the Golden Rule—that embodies fairness in all human societies: "Do unto others as you would have them do unto you." It has guided the course of history, shaped society, and ruled our social lives. At the end of our journey through this book, we have encountered only a small set of examples. But the message should be abundantly clear—our fairness instinct is supremely powerful.

Having emerged as a necessary element of social living and evolved under both biological and cultural pressures over time, the fairness instinct governs our social interactions and holds us together as a society. The instinct can be so overwhelming that, if unsatisfied, change becomes inevitable and can happen at any cost. This is why couples argue, fight, and divorce; friends turn against each other; families and neighbors feud; coworkers and colleagues become nemeses; political parties bicker; religions split; nations divide; countries go to war; revolutions surge; terrorists strike; empires fall. So while the form of any particular society may be unimportant, the function of all societies must eventually adapt to our fairness instinct. "Fairness is like a 'golden thread' that binds together a harmonious society," writes social thinker Peter Corning. "And when this thread breaks, the social fabric unravels."[1] In this sense, fairness is a society's way of making another society—a society that better fits our fairness instinct.

The quintessential appeal of democracy is the principle of equality for all. This has been unmistakably a leitmotif of the Arab Spring and vagaries of protests in many other nations. In any established democratic society, be it in North America, Europe, or elsewhere, part of the promise of fairness is not that we all start off or end up equal, but that we have equal opportunity to rise according to our own merit, choices, and effort. But when inheritance reemerges as a predictor of a person's success, social mobility snaps closed. When one's state of birth matters more than one's intelligence, ambition, motivation, or effort, the fundamental backbone of democracy—the principle of equality for all—is broken. Churchill's tongue-in-cheek remark—"Democracy is the worst form of government, except for all those other forms that have been tried from time to time"[2]—may have to be revised.

I have drawn many examples from China in this book. China's economy has been expanding with unprecedented speed and scale, and, as a result, it is the theater where the most exciting show is playing out today. And under the spotlight is an underappreciated opportunity to observe the human fairness instinct asserting its influence, in real time as a society evolves. As China basks in economic growth, rising in just two decades from a poor and backwater Communist country to become a rich, dynamic, and vibrant economic powerhouse, we can observe the lion waking up, fulfilling Napoleon's famous prediction made two centuries earlier. However, as the number of Chinese billionaires exceeded one hundred in 2012, the specter of soaring economic inequality has begun to haunt the nation, bringing instability and a battery of serious social problems that exhibit a familiar scenario, one experienced by many other expanding economies over time. In a malfunctioning society, observes journalist Alan Honick, "inequality is the symptom; unfairness is the underlying disease."[3] In this regard, China is not unique. The same disease naturally shows similar symptoms in other societies.

Though its mature economy is experiencing markedly slower growth than China's rapid expansion, the United States, too, has become increasingly mired in a sea of popular discontent, with increasing economic inequality, resulting in social and political unrest. From the Tea Party complaints to the protests of Occupy Wall Street, the public ire with regard to the status quo has built up to what many believe is a tipping point. A major change may loom on the horizon. This is not intended as a doomsday prediction, however. Change need not be revolutionary, destroying the existing order to start anew. On the contrary, it can be evolutionary, ushering in a gradual, smooth transition to a society that may better satisfy the fairness instinct in more of us. The good news for now is that we still have the choice of which path to take.

ACKNOWLEDGMENTS

The idea for this book was conceived over a decade ago. Its long-delayed birth was due to two daunting challenges: the limited scope of my knowledge, and my lack of confidence in my ability to write in English for a general audience. Fortunately, many have helped me overcome my self-doubt. *The Fairness Instinct* represents a collective intellectual journey I have traveled with them.

David Barash has been an inspirational figure. In addition to providing encouragement, he also made many insightful suggestions. Marte Fallshore was an invaluable resource, especially for issues related to psychology as well as English. Lisa Norris generously took me as a student in her writing courses. Stephen Maurer, Elizabeth Maurer, Ariel Knafo, John Dunlop, Eric Dolin, Charles Li, Erik Borst, Debbie Lewis, and Amy Zhang all contributed in various ways. Several staff members at the library of Central Washington University helped me to obtain reference books and articles. *Wikipedia* was a constant companion, providing leads for further research and making it much easier to synthesize information from vastly different domains of knowledge.

My heartfelt thanks go to Andrew Willden, whose contributions were comprehensive, from grammar, rhetoric, readability, and book organization to information about religion and history. His expertise in Mormonism and ancient Western history was essential in helping me avoid basic mistakes. His superb editing skill has considerably improved the prose and enriched the context of several historical episodes. I am equally grateful to Alan Honick, who passionately edited the entire manuscript in addition to engaging with me in many hours of discussion. The book would have been less readable without his contribution. Another major and unfailing helper was my son, Shine. Unusually mature in writing and editing skills for a middle school student, Shine was always present throughout my writing process, and he offered countless insights on issues ranging from wording, grammar, and rhetoric to logic, flow, and the structure and organization of every chapter from rough draft to final version.

My sincere thanks also go to my agent Don Fehr, who spotted the merit of the project and guided me through to the finish line. It has also been a wonderful experience working with Prometheus Books. Efficient, enthusiastic, and always encouraging, Brian McMahon is every author's dream editor, and I am lucky to have had him as mine. Many others, including Steven Mitchell, Melissa Shofner, Cate Roberts-Abel, Mariel Bard, and Mark Hall, also helped immeasurably in bringing out this book.

Finally, my wife Crystal willingly played devil's advocate, relentlessly probing the potential weaknesses in my writing. This book was much improved as a result of my effort to meet her high standards.

Because many ideas in the book are new, they are yet to be further examined and criticized. I take full responsibility for any mistakes that may be found. I believe that the issues surrounding fairness are extremely important and that the risk of ignoring them is extremely grave; for this reason, I am willing to risk my reputation to the gnashing teeth of the critics. As the old saying goes: "throw out a minnow to catch a whale." In this sense, I will be satisfied if this book fulfills its role as a bait minnow to bring in a bigger catch: an earnest, informed discussion of how societies may be reformed to satisfy the basic human instinct for fairness.

NOTES

PART I: THE ROOTS OF THE ROBIN HOOD MENTALITY

Chapter 1: Introduction, or, Who Is Robin Hood?

1. M. Keen, *The Outlaws of Medieval Legend* (London: Routledge & Kegan Paul, 1977), p. xvi.

2. "List of Television Series Featuring Robin Hood," *Wikipedia*, shttp://en.wikipedia.org/wiki/List_of_films_and_television_series_featuring_Robin_Hood (accessed February 10, 2012).

3. The Robin Hood Project, University of Rochester, http://www.lib.rochester.edu/camelot/rh/rhhome.htm (accessed May 26, 2011).

4. Keen, *Outlaws of Medieval Legend*, pp. 9–77.

5. "Robin Hood," *Wikipedia*, http://en.wikipedia.org/wiki/Robin_Hood#Popular_culture (accessed February 10, 2012). For other dates that were mentioned by various authors, see J. C. Holt, *Robin Hood* (London: Thames & Hudson, 1982), p. 51.

6. Holt, *Robin Hood*, p. 41. This date does not exist in the Roman calendar. However, according to M. Parker's *The True Tale of Robin Hood* (1632), Robin Hood died on December 4, 1198. This, if true, would exclude the possibility that Robin Hood was active in 1228.

7. Keen, *Outlaws of Medieval Legend*, pp. 176–77.

8. Ibid., 175–76.

9. Yeoman could also refer to a person who worked for an aristocrat.

10. J. Bellamy, *Robin Hood: An Historical Enquiry* (Bloomington: Indiana University Press, 1985), p. 132.

11. Holt, *Robin Hood*, pp. 7–8.

12. Keen, *Outlaws of Medieval Legend*, p. 191.

13. J. Ritson also agreed with Robin Hood's noble identity (see Bellamy, *Robin Hood*, pp. 2–3).

14. Holt, *Robin Hood*, p. 42.

15. Ibid., p. 9.

16. Ibid., p. 7.

17. For more on European folk heroes, see T. Blanning, *The Pursuit of Glory: The Five Revolutions That Made Modern Europe: 1648–1815* (New York: Penguin, 2007), pp. 174–80.

18. Holt, *Robin Hood*, p. 10.

19. Cited from R. C. Solomon, *A Passion for Justice: Emotions and the Origins of the Social Contract* (Reading: Addison-Wesley, 1990), p. 4.

20. D. Schmidtz, *Elements of Justice* (Cambridge: Cambridge University Press, 2006), p. 7.

21. A. Sen, *The Idea of Justice* (Cambridge, MA: Harvard University Press, 2009), p. xv.

22. S. Fleischacker, *A Short History of Distributive Justice* (Cambridge, MA: Harvard University Press, 2004), pp. 2, 5, 10, 19. It was Aristotle who made a distinction between distributive justice, which deals with money and political office to be allocated according to merit, and corrective justice (later known as commutative justice), which deals with wrongdoers' compensation for the damage they have caused to their victims.

23. Aristotle, *The Nicomachean Ethics* (Oxford: Oxford University Press, 2009), p. 85.

24. Fleischacker, *Short History of Distributive Justice*, p. 6.

25. D. Johnston, *A Brief History of Justice* (Chichester, Wiley-Blackwell, 2011), p. 95.

26. Fleischacker, *Short History of Distributive Justice*, pp. 21–22.

27. See ibid., pp. 68–75, for more details about Kant's idea of distributive justice.

28. J. Locke, *Two Treatises on Government* (London: Butler, 1821), p. 46.

29. D. Hume, *A Treatise on Human Nature* (London: Longmans, Green, 1878), p. 256.

30. Fleischacker, *Short History of Distributive Justice*, p. 40.

31. Ibid., p. 43.

32. Ibid., pp. 5, 7. Fleischacker provides at least five premises about the modern concept of distributive justice.

33. Rawls also developed and advocated for the "veil of ignorance" idea for considering the initial position in moral judgment. The idea is a thought experiment where, when judging what is fair and just, a person should be completely detached—veiled—from his or her talents, preferences, and socioeconomic standing. This idea was initially conceived by economist John Harsanyi.

34. J. Rawls, *Justice as Fairness: A Restatement* (Cambridge, MA: Harvard University Press, 2001), pp. 14–15.

35. See E. Kelly, "Forward," in *Justice as Fairness*, p. xi.

36. Sen, *Idea of Justice*, p. 54.

37. Psychological studies of moral issues such as justice were pioneered by J. Piaget. But at the beginning they were narrowly defined and approached from the developmental and learning perspectives in children without reference to the genetic perspective. Only in the 1990s did we begin to see major attempts by scientists to fill the gap.

38. J.-J. Rousseau, *Discourse on Inequality* (Whitefish: Kessinger, 1754/2004), p. 13.

39. Solomon, *Passion for Justice*, p. 3.

40. M. D. Hauser (*Moral Minds: How Nature Designed Our Universal Sense of Right and Wrong* [New York: HarperCollins, 2006], p. 31) provides a distinction between intuitions and conscious reasoning: "Intuitions are fast, automatic, involuntary, require little attention,

appear early in development, are delivered in the absence of principled reasons, and often appear immune to counter-reasoning. Principled reasoning is slow, deliberate, thoughtful, requires considerable attention, appears late in development, [is] justifiable, and [is] open to carefully defended and principled counterclaims."

41. In his article in the *Dictionary of National Biography*, historian S. Lee writes of his belief that Robin Hood never existed as a person (Holt, *Robin Hood*, pp. 7–8, 40).

42. Keen, *Outlaws of Medieval Legend*, p. 214.

43. Ibid., p. 6.

44. Ibid., p. 209.

45. Ibid., p. 145.

Chapter 2: The Making of Robin Hood

1. R. D. Alexander, *The Biology of the Moral Systems* (New York: Aldine de Gruyter, 1987), p. xiv.

2. For a plain explanation, see "Occam's Razor," *Wikipedia*, http://en.wikipedia.org/wiki/Occam%27s_razor (accessed February 8, 2012).

3. His original claim in Latin reads, "*entia non sunt multiplicanda praeter necessitate*."

4. Elegance here connotes, as some contemporary science philosophers believe, that parsimonious hypotheses are those that are clearly stated and easy to test.

5. This consideration excludes situations in other social mammals such as wolves, dolphins, and whales, whose sense of fairness is not as evident as that in primates.

6. In Brosnan and de Waal's study, although only quality of reward (grape versus cucumber) was involved, the *sizes* of the rewards were controlled to be approximately the same. Though not tested, it is possible that the capuchin monkeys might accept two or more slices of cucumber when their peers obtain one grape. This was confirmed in a similar study, where trained capuchin monkeys responded by "buying" more of a cheaper food reward and less of a more expensive food reward with tokens (M. K. Chen, V. Lakshminarayanan and L. R. Santos, "How Basic Are Behavioral Biases? Evidence from Capuchin Monkey Trading Behavior," *Journal of Political Economy* 114 [2006]: 517–37).

7. For the butterfly, see D. J. Kemp, "Female Mating Biases for Bright Ultraviolet Iridescence in the Butterfly *Eurema hecabe* (Pieridae)," *Behavioral Ecology* 19 (2007): 1–8; for the guppy, see A. E. Houde and J. A. Endler, "Correlated Evolution of Female Mating Preferences and Male Color Patterns in the Guppy *Poecilia reticulata*," *Science* 248 (1990): 1405–1408; for the Túngara frog, see M. J. Ryan and A. S. Rand, "Sexual Selection and Signal Evolution: Ghost of Biases Past," *Philosophical Transactions of the Royal Society of London* B 340 (1993): 187–95; for the peacock, see A. Loyau et al., "Multiple Sexual Advertisements Honestly Reflect Health Status in Peacocks (*Pavo cristatus*)," *Behavioral Ecology and Sociobiology* 58 (2005): 552–57.

8. Birds that can be stimulated to replace removed eggs are called indeterminate egg

layers whereas those that do not respond to such manipulation are called determinate egg layers.

9. B. E. Lyon, "Egg Recognition and Counting Reduce Costs of Avian Conspecific Brood Parasitism," *Nature* 422 (2003): 495–99.

10. E. M. Brannon, "What Animals Know about Numbers," in *Handbook of Mathematical Cognition*, ed. J. I. D. Campbell (New York: Psychology Press, 2005), pp. 85–107.

11. K. C. Berridge and T. E. Robinson, "What Is the Role of Dopamine in Reward: Hedonic Impact, Reward Learning, or Incentive Salience?" *Brain Research Review* 28 (1998): 309–69; W. Schultz, "Predictive Reward Signal of Dopamine Neurons," *Journal of Neurophysiology* 80 (1998): 1–27; W. Schultz, "Dopamine Signals for Reward Value and Risk: Basic and Recent Data," *Behavioral and Brain Functions* 6 (2010): 24–32.

12. E. S. Bromberg-Martin, M. Matsumoto, and O. Hikosaka, "Dopamine in Motivational Control: Rewarding, Aversive, and Alerting," *Neuron* 68 (2010): 815–34; Schultz, "Dopamine Signals for Reward Value and Risk."

13. In humans, an innate deficiency of empathy may occasionally lead to nightmarish personalities typified by psychopaths, a label that brings to mind such infamous American serial killers as Ted Bundy, Gary Ridgeway, and Andrew Cunanan. Even a slight reduction of empathy may bring unpleasant consequences. Some people with antisocial tendencies can be doggedly unpleasant—often dubbed as a "pain in the neck" in the workplace—due to their poor understanding of, and inept response to, their colleagues' emotions. The primary reason may not be attributable to their apparent apathy or rudeness but to their deficient ability to read others' feelings.

14. S. Blakeslee, "Cells That Read Minds," *New York Times*, January 10, 2006. More non-technical information about mirror neurons and spindle cells can be found in M. Bekoff and J. Pierce, *Wild Justice* (Chicago: University of Chicago Press, 2009), pp. 29–30, 158.

15. G. Rizzolatti, L. Fogassi and V. Gallese, "Neurophysiological Mechanisms Underlying the Understanding and Imitation of Action," *Nature Review Neuroscience* 2 (2001): 661–70; D. E. Lyons, L. R. Santos and F. C. Keil, "Reflections of Other Minds: How Primate Social Cognition Can Inform the Function of Mirror Neurons," *Current Opinion in Neurobiology* 16 (2006): 230–34.

16. Bromberg-Martin, Matsumoto, and Hikosaka, "Dopamine in Motivational Control."

17. P. R. Hof and E. van der Gucht, "Structure of the Cerebral Cortex of the Humpback Whale, *Megaptera novaeangliae* (Cetacea, Mysticeti, Balaenopteridae)," *Anatomical Record* 290 (2006): 1–31.

18. There is no absolutely solitary species in sexually reproducing animals. At the very least, sexual partners must interact for mating. Here we use "social living" to refer to animals that stay in relatively stable groups, in which at least some, if not all, members recognize one another.

19. W. D. Hamilton, "The Genetical Evolution of Social Behavior," *Journal of Theoretical Biology* 7 (1964): 1–52.

20. Women were married out, and their kinship statuses were redefined in their husbands'

families or clans. See L. Sun, "Blood Is Not Always Thicker than Water: The Limited Effect of Kin Selection on Human Kinship in the Traditional Chinese Family," *Current Zoology* 56 (2010): 182–89.

21. Ibid.

22. R. L. Trivers, "The Evolution of Reciprocal Altruism," *Quarterly Review of Biology* 46 (1971): 35–57.

23. See G. Hardin, "Discriminating Altruisms," *Zygon* 17 (1982): 163–86.

24. G. S. Wilkinson, "Food Sharing in Vampire Bats," *Scientific American* 262, no. 2 (1990): 64–70.

25. In the United States, people are willing to pay a fifth of GDP for maintaining distributive justice (see G. Corneo and C. M. Fong, "What's the Monetary Value of Distributive Justice?" CESIFO Working Paper, no. 1706 [2006]).

26. R. Noë and P. Hammerstein, "Biological Markets: Supply and Demand Determine the Effect of Partner Choice in Cooperation, Mutualism, and Mating," *Behavioral Ecology and Sociobiology* 35 (1994): 1–11; R. Noë and P. Hammerstein, "Biological Markets," *TREE* 10 (1995): 336–39; R. Noë, "Digging for the Roots of Trading," in *Cooperation in Primates and Humans*, ed. P. M. Kappeler and C. P. van Schaik (Berlin: Springer, 2006), pp. 233–61.

27. For the chacma baboon, see L. Barrett et al., "Market Forces Predict Grooming Reciprocity in Female Baboons," *Proceedings of the Royal Society of London* B 266 (1999): 665–70; for the Tibetan macaque, see D. Xia et al., "Grooming Reciprocity in Female Tibetan Macaques *Macaca thibetana*," *American Journal of Primatology* 74 (2012): 569–79.

28. M. D. Gumert, "Payment for Sex in a Macaque Mating Market," *Animal Behaviour* 74 (2007): 1655–67.

29. I. Norscia, D. Antonacci, and E. Palagi, "Mating First, Mating More: Biological Market Fluctuation in a Wild Prosimian," *Plos ONE* 4, no. 3 (2009): 1–6.

30. R. Thornhill, "Sexual Selection in the Black-Tipped Hangingfly," *Scientific American* 242, no. 6 (1980): 162–72.

31. For those who are uncomfortable with haggling, fair trade is actually quite easy to accomplish, even when facing a shrewd bargainer. The trick, as a mathematical model shows, is to stick to your position in trade negotiation. See T. Ellingsen, "The Evolution of Bargaining Behavior," *Quarterly Journal of Economics* 112 (1997): 581–602.

32. V. Lakshminaryanan, M. K. Chen, and L. R. Santos, "Endowment Effect in Capuchin Monkeys," *Philosophical Transactions of the Royal Society* 363 (2008): 3837–44; L. R. Santos and M. K. Chen, "The Evolution of Rational and Irrational Economic Behavior: Evidence and Insight from a Non-Human Primate Species," in *Neuroeconomics: Decision Making and the Brain*, ed. P. W. Glimcher et al. (London: Elsevier, Academic Press, 2009), pp. 81–93.

33. In nonhuman primates such as baboons and chimpanzees, lower ranking individuals often form coalitions and alliances that can effectively fight against higher ranking individuals. This can make a social hierarchy more dynamic and despotism more difficult.

34. This bottom-up force for leveling social status in a hierarchy, though never explicitly addressed, is implied in several scholarly works such as those by C. Boehm, *Hierarchy in the Forest* (Cambridge, MA: Harvard University Press, 1999), and R. Folger and R. Cropanzano, "Social Hierarchies and the Evolution of Moral Emotions," in *Managerial Ethics: Managing the Psychology of Morality*, ed. M. Schminke (New York: Routledge, 2010), pp. 207–34.

35. J. Tata, "Influence of Role and Gender on the Use of Distributive versus Procedural Justice Principles," *Journal of Psychology* 134 (2000): 261–68.

36. C. B. Burgoyne and A. Lewis, "Distributive Justice in Marriage: Equality or Equity?" *Journal of Community and Applied Social Psychology* 4 (1994): 101–14.

37. Churchill made this point in his speech to British Parliament on May 13, 1940, just three days after he became prime minister.

38. K. M. Page and M. A. Nowak, "Empathy Leads to Fairness," *Bulletin of Mathematical Biology* 64 (2002): 1101–16.

39. A fairness mentality based on the possibility of table turning was nicely illustrated in the wake of the National Football League game between the Seattle Seahawks and the Green Bay Packers on September 24, 2012. A last second missed call by poorly trained replacement referees handed the Seahawks an almost impossible win. While the Packers' fans were aggrieved, the jubilant Seahawks fans also wanted competent, unbiased referees in the future because it would be possible for their team to suffer next time.

Chapter 3: Robin Hood in Our DNA

1. For the ease of remembering, the "P" in "Peter" stands for the "P" in "Proposer," and the "R" in "Richard" for the "R" in "Recipient."

2. The amount of money involved in the game is normally not an issue, unless it is large, such as a few hundred dollars. When the sum is large, the rate of rejection is slightly lower. For more detail, see C. F. Camerer, *Behavioral Game Theory: Experiments in Strategic Interaction* (Princeton, NJ: Princeton University Press, 2003), pp. 60–63.

3. The Ultimatum Game is not entirely hypothetical. Camerer (*Behavioral Game Theory*, p. 44) shows a real-world case in American history, during the Constitutional Convention of 1887. Delegates debated the proposal by Pennsylvania delegate Gouverneur Morris as to whether, in the expansion of the original thirteen states, the new western states should be admitted to the Union as second-rate states. George Mason, a Virginia statesman, instantly sensed the possibility that the western states might reject the offer of joining the Union if they were granted an inferior status: "They [people in the western states] will have the same pride and other passions which we have," he argued, "and will either not unite with or will speedily revolt from the Union, if they are not in all respects placed on an equal footing with their brethren."

4. One way to meet these stringent conditions is to prevent the players from seeing each other so that there is no way for the players to know with whom they are playing.

5. See the summary in Camerer's book (*Behavioral Game Theory*, pp. 48–59). Data presented in the next paragraph are also from this source.

6. D. Kahneman, J. L. Knetsch and R. Thaler, "Fairness as a Constraint on Profit Seeking: Entitlements in the Market," *American Economic Review* 76 (1986): 728–41.

7. R. Forsythe et al., "Fairness in Simple Bargaining Experiments," *Games and Economic Behavior* 6 (1994): 347–69.

8. M. A. Nowak and K. Sigmund, "Evolution of Indirect Reciprocity by Image Scoring," *Nature* 393 (1998): 573–77; M. A. Nowak and K. Sigmund, "Evolution of Indirect Reciprocity," *Nature* 437 (2005): 1291–98. This form of cooperation based on reputation is called indirect reciprocity.

9. Filial piety was perpetrated often at the expense of the well-being of younger generations when food and money were strained. Sometimes, younger family members sacrificed their comfort, health, or career opportunities to serve their aging parents or grandparents. Superficial analysis may indicate that it should be the young and strong, with great reproductive value ahead of them, who should be favored, not the elders. It is indeed the case in many other societies: family elders often dedicate themselves to younger generations, a phenomenon dubbed "the grandma effect" by behavioral scientists. The Chinese notion of filial piety is entirely missing in the West, for which philosopher B. Russell (*The Problem of China* [London: George Allen & Unwin, 1922], p. 40) was deeply puzzled, as it "departs seriously from common sense."

10. Dynastic laws in ancient China also provided benefits to filial people and punished those who dodged filial duties.

11. G. Tabibnia and M. D. Lieberman, "Fairness and Cooperation Are Rewarding: Evidence from Social Cognitive Neuroscience," *Annals of New York Academy of Sciences* 1118 (2007): 90–101.

12. Rational choice theory is highly influential in microeconomics. G. Becker, for instance, won the 1992 Nobel Prize in Economics for his development of rational choice models in education, crime, and discrimination. (For more detail, see "Rational Choice Theory," *Wikipedia*, http://en.wikipedia.org/wiki/Rational_choice_theory [accessed February 20, 2012]).

13. W. T. Harbaugh, K. Kraus, and S. G. Liday, "Bargaining by Children," University of Oregon Economics Department Working Paper 2003 (available at http: //economics.uoregon .edu [accessed February 15, 2012]). J. K. Murnighan and M. S. Saxon ("Ultimatum Bargaining by Children and Adults," *Journal of Economic Psychology* 19 [1998]: 415–45) obtained almost the same results in their study of third and sixth graders.

14. E. Fehr, H. Bernhard, and B. Rockenbach, "Egalitarianism in Young Children," *Nature* 454 (2008): 1079–84.

15. Ibid. E. Fehr and his colleagues have confirmed that the transition from acting on gut feelings to egalitarianism in children happens between the ages of three and eight.

16. B. Wallace et al., "Heritability of Ultimatum Game Responder Behavior," *Proceedings of National Academy of Sciences, USA* 104 (2007): 15631–34.

17. For a mathematical model showing that rejecting low offers can boost reputation, see M. A. Nowak, K. M. Page, and K. Sigmund, "Fairness versus Reason in the Ultimatum Game," *Science* 289 (2000): 1773–75.

18. A. G. Sanfey et al., "The Neural Basis of Economic Decision-Making in the Ultimatum Game," *Science* 300 (2003): 1755–58.

19. D. Knoch et al., "Diminishing Reciprocal Fairness by Disrupting the Right Prefrontal Cortex," *Science* 314 (2006): 829–32.

20. Evolutionary biologists distinguish between proximate causations and ultimate causations. When we are hungry, for instance, the causes can be an empty stomach, a low blood-sugar level, a higher than normal metabolic rate, or the mere presence of delicious food. These are all proximate causations, answering *how* the sensation of hunger is incurred. The ultimate causation of hunger deals with *why* the sensation of hunger has been endowed to us by evolution. A person who has no sense of hunger (an anorexic patient, for instance) will suffer a lower level of fitness and will be selected against. Accordingly, I use "relative fitness" to refer to ultimate causations and "relative payoff" to proximate causations in the book.

21. P. R. Grant, *Ecology and Evolution of Darwin's Finches* (Princeton, NJ: Princeton University Press, 1986); B. R. Grant and P. R. Grant, *Evolutionary Dynamics of a Natural Population: The Large Cactus Finch of the Galápagos* (Chicago: University of Chicago Press, 1989); P. R. Grant and B. R. Grant, "Unpredictable Evolution in a 30-Year Study of Darwin's Finches," *Science* 296 (2002): 707–11.

22. In a two-person game, any difference can make one better off than the other, for which any unequal offer should be rejected based on relative payoff, as illustrated in the hypothetical monkey case. However, this is unrealistic because evolution takes place in the entire group, not the two players isolated from the rest. So Richard has to compare his gain with Peter's as well as those who are not directly involved in the game. This may be the evolutionary reason why he would only reject a certain level of difference but not all differences.

23. Sanfey et al., "Neural Basis of Economic Decision-Making in the Ultimatum Game."

24. M. D. Alicke and E. Zell, "Social Comparison and Envy," in *Envy: Theory and Research*, ed. R. H. Smith (Oxford: Oxford University Press, 2008), pp. 73–93.

25. For Florida, see J. A. List and T. L. Cherry, "Learning to Accept in Ultimatum Games: Evidence from an Experimental Design that Generates Low Offers," *Experimental Economics* 3 (2000): 11–31; for the Slovak Republic, see R. Slonim and A. E. Roth, "Learning in High Stakes Ultimatum Games: An Experiment in the Slovak Republic," *Econometrica* 66 (1998): 569–96; for Indonesia, see L. A. Cameron, "Raising the Stakes in the Ultimatum Game: Experimental Evidence from Indonesia," *Economic Inquiry* 27 (1999): 47–59.

26. J. Henrich et al., "In Search of *Homo economicus*: Behavioral Experiments in 15 Small-Scale Societies," *American Economic Review* 91 (2001): 73–78.

27. Studies at the Keatley Creek settlement in British Columbia by archeologist B. Hayden show that competitive feasting, besides reputation enhancing, was a way of creating

debt relationships, gaining power, and ultimately redefining the norms of fairness in some pre-industrial societies. See A. Honick, "The Evolution of Fairness," *Pacific Standard*, August 31, 2012 (available at http://www.psmag.com/culture/the-evolution-of-fairness-45681/ (accessed February 10, 2013).

28. The equity principle of justice says that individuals who invest the most should receive the most whereas the equality principle of justice mandates that all individuals should receive equal shares of a fortune. See C. W. Leach, "Envy, Inferiority, and Injustice: Three Bases of Anger about Inequality," in *Envy: Theory and Research*, ed. R. H. Smith (Oxford: Oxford University Press, 2008), pp. 94–116.

29. P. Kaplan, P. P. Wang, and U. Francke, "Williams (Williams Beuren) Syndrome: A Distinct Neurobehavioral Disorder," *Journal of Child Neurology* 16 (2001): 177–90; A. Meyer-Lindenberg, C. B. Mervis, and K. F. Berman, "Neural Mechanisms in Williams Syndrome: A Unique Window to Genetic Influences on Cognition and Behavior," *Nature Reviews/Neuroscience* 7 (2006): 380–93; A. Järvinen-Pasley et al., "Defining the Social Phenotype in Williams Syndrome: A Model for Linking Gene, the Brain, and Behavior," *Developmental Psychopathology* 20 (2008): 1–35; M. A. Martens, S. J. Wilson, and D. C. Reutens, "Research Review: Williams Syndrome: A Critical Review of the Cognitive, Behavioral, and Neuroanatomical Phenotype," *Journal of Child Psychology and Psychiatry* 49 (2008): 576–608.

30. J. H. Dulebohn et al., "The Biological Bases of Unfairness: Neuroimaging Evidence for the Distinctiveness of Procedural and Distributive Justice," *Organizational Behavior and Human Decision Processes* 110 (2009): 140–51.

31. B. Güroğlu et al., "Unfair? It Depends: Neural Correlates of Fairness in Social Context," *Scan* 5: 414–23.

32. E. Tricomi et al., "Neural Evidence for Inequality-Averse Social Preferences," *Nature* 463 (2010): 1089–91.

33. E. S. Bromberg-Martin, M. Matsumoto, and O. Hikosaka, "Dopamine in Motivational Control: Rewarding, Aversive, and Alerting," *Neuron* 68 (2010): 815–34.

34. R. Avinun et al., "AVPR1A Variant Associated with Preschoolers' Lower Altruistic Behavior," *Plos ONE* 6, no. 9 (2011): 1–5. The expression pattern of the *AVPR1A* gene appears to be age specific or brain region specific, as different results have also been obtained (see A. Knafo et al., "Individual Differences in Allocation of Funds in the Dictator Game Associated with Length of the Arginine Vasopressin 1a Receptor RS3 Promoter Region and Correlation between RS3 Length and Hippocampal mRNA," *Genes, Brain and Behavior* 7 [2008]: 266–75).

35. E. Emanuele et al., "Relationship between Platelet Serotonin Content and Rejections of Unfair Offers in the Ultimatum Game," *Neuroscience Letters* 437 (2008): 158–61.

36. S. Chew, R. Ebstein and S. Zhong, "Sex-Hormone Genes and Gender Difference in Ultimatum Game: Experimental Evidence from China and Israel," December 22, 2011 (available at http://papers.ssrn.com/sol3/papers.cfm?abstract_id=1975735, accessed March 12, 2013).

37. S. Zhong et al., "Dopamine D4 Receptor Gene Associated with Fairness Preference in Ultimatum Game," *Plos ONE* 5, no. 11 (2010): 1–8.

38. This principle of distribution captures the gist of the classical equality theory. See J. S. Adams, "Toward an Understanding of Inequity," *Journal of Abnormal and Social Inequity* 67 (1963): 422–36; J. S. Adams, "Inequity in Social Exchange," in *Advances in Experimental Social Psychology*, vol. 2, ed. L. Berkowitz (New York: Academic Press, 1965), pp. 267–99; M. E. Price, "Free Riders as a Blind Spot of Equity Theory: An Evolutionary Correction," in *Managerial Ethics: Managing the Psychology of Morality*, ed. M. Schminke (New York: Routledge, 2010), pp. 235–56.

39. S. F. Brosnan, H. C. Schiff, and F. B. M. de Waal, "Tolerance for Inequity May Increase with Social Closeness in Chimpanzees," *Proceedings of the Royal Society of London* B 272 (2005): 253–58; M. van Wokenten, S. F. Brosnan, and F. B. M. de Waal, "Inequity Responses of Monkeys Modified by Effort," *Proceedings of National Academy of Sciences, USA* 104 (2007): 18854–59.

40. See letter dated March 14, 2009, from Edward M. Liddy, Chairman and Chief Executive Officer of AIG, to Timothy F. Geithner, United States Secretary of the Treasury, at Clips & Comment, http://clipsandcomment.com/documents/liddy.pdf (accessed June 26, 2013).

41. M. J. Sandel, *Justice: What's the Right Thing to Do?* (New York: Farrar, Straus & Giroux, 2010), p. 15.

42. A British equivalent of the AIG controversy was the case of Fred Goodwin, whose leadership at the Royal Bank of Scotland resulted in the loss of £24.1 billion, the largest annual corporate loss in UK history. Still, he retired with a very generous pension of about £700,000, which stirred up a broad outrage in Britain. See Price, "Free Riders as a Blind Spot of Equity Theory," pp. 235–56.

43. Some Americans, especially conservatives, are also against labor unions, which tend to be popular among autoworkers.

44. See "State of the Union: President Obama's Speech," ABC News, http://abcnews.go.com/Politics/State_of_the_Union/state-of-the-union-2010-president-obama-speech-transcript/ story?id=9678572 (accessed June 26, 2013).

PART 2: THE SURVIVAL OF THE MEDIOCRE

Chapter 4: Life in a "Red-Eyed" Commune

1. Jane's Chinese name is Yajuan, which isn't easy to pronounce for English speakers.

2. The growth season is about four weeks short of double cropping for rice in the region. To make up this shortage in time, harvesting and planting need to be done at the same time.

3. Fortunately, things were extremely cheap, and peasants could get by with little. Five

yuan (about US$0.70) could cover a semester's tuition, books, and other expenses for a school child. Seventeen yuan (about two dollars) could buy a nice violin. An elementary teacher's monthly salary was typically twenty-some yuan, or three dollars. Ten cents (0.7 yuan) could buy a pound of pork, and three cents (0.2 yuan), a pound of wild yellow croakers (a popular seafood in the coastal area of the East China Sea, where Jane's community was located), which now sell for about US$200.

4. Surprisingly, envy in several cultures around the Mediterranean is also compared to "red eyes," which, likewise, have the magical power to cause misfortune to the envied (see C. Maloney, ed., *The Evil Eye* [New York: Columbia University Press, 1976]). Here, cultures, enmeshed in human nature, often coincide spectacularly.

5. Though similar in emotion, jealousy and envy are different in connotation. Jealousy refers to human relationships, especially under the context of mate competition.

Chapter 5: The Survival of the Mediocre

1. E. Fehr and S. Gächter, "Altruistic Punishment in Humans," *Nature* 415 (2002): 137–40.

2. Fehr and Gächter used a generic currency, which they called the Money Unit, for the experiment. One Money Unit was about one US dollar.

3. This game is known in game theory as the Public Goods Game.

4. This value of 60 percent was strategically chosen so that if there was only one person who invested, his or her gain would never exceed 100 percent, regardless of the increment in the team fund.

5. The calculation is as follows: 4 players × $20 per player = $80 invested in the team fund, which can grow to be $80 × $0.4 = $32 for each player.

6. K. Binmore, *Natural Justice* (Oxford: Oxford University Press, 2005), p. 89.

7. For a more detailed account of how the blind fish evolved, see T. E. Dowling, D. P. Martasian, and W. R. Jeffery, "Evidence for Multiple Genetic Forms with Similar Eyeless Phenotypes in the Blind Cavefish, *Astyanax mexicanus*," *Molecular Biology and Evolution* 19 (2002): 446–55.

8. C. Boehm, *Moral Origins: The Evolution of Virtue, Altruism, and Shame* (New York: Basic Books, 2012), p. 107.

9. Experimental studies show that people who play the role of a third party take actions based on selfish motives. See, for example, A. Leibbrandt and R. López-Pérez, "An Exploration of Third and Second Party Punishment in Ten Simple Games" (available at http://www.uam.es/personal_pdi/economicas/ralopez/JEBO%20with%20appendix.pdf [accessed February 11, 2013]).

10. A. K. Choy, T. D. Fields, and R. R. King, "Social Responsibility of the Auditor as Independent Gatekeeper: An Experimental Investigation" (available at http://dx.doi.org/10.2139/ssrn.1248342 [accessed February 11, 2013]). Such a game is called the Gatekeeper

Game, a modification of the Ultimatum Game with the voting power transferred to the third party, the gatekeeper.

11. There are several slightly different versions for the metaphor of the "hedonic treadmill." The one here is closest to psychologist M. Eysenck's notion in his book *Happiness: Facts and Myths* (Hove: Lawrence Erlbaum, 1990).

12. See, for example, R. H. Frank, *Luxury Fever: Why Money Fails to Satisfy in an Era of Excess* (New York: Free Press, 1999). In fact, people whose incomes are well above the average may also experience the same feeling when compared with better-off peers.

13. S. E. Hill and D. M. Buss, "Envy and Positional Bias in the Evolutionary Psychology of Management," *Managerial and Decision Economics* 27 (2006): 131–43.

14. F. Bacon, *The Works of Francis Bacon* (London: Millar, 1765), p. 475.

15. I. Kant, *The Metaphysics of Morals* (Cambridge: Cambridge University Press, 1797/2000), p. 459.

16. R. H. Smith and S. H. Kim, "Introduction," in *Envy: Theory and Research*, ed. R. H. Smith (Oxford: Oxford University Press, 2008), pp. 1–14.

17. Ibid. Smith and Kim distinguish between psychological envy, which is often based on perceived injustice, and resentment, which, in traditional philosophy, is a moral sentiment.

18. A. Schopenhauer, *The Pessimist's Handbook: A Collection of Popular Essays* (Lincoln: University of Nebraska Press, 1976), p. 346.

19. B. Vidaillet, *Workplace Envy* (New York: Palgrave Macmillan, 2008), p. 35.

20. D. Hume, *A Treatise of Human Nature* (Sioux Falls, SD: NuVision, 1739/2007), p. 271.

21. B. Spinoza, *Ethics* (Hertfordshire: Wordsworth, 1677/2001), p. 140.

22. W. G. Parrott and R. H. Smith, "Distinguishing the Experiences of Envy and Jealousy," *Journal of Personality and Social Psychology* 64 (1993): 906–20.

23. S. A. Bers and J. Rodin, "Social-Comparison Jealousy: A Developmental and Motivational Study," *Journal of Personality and Social Psychology* 47 (1984): 766–79.

24. Modified from J. E. Joseph et al., "Functional Neuroanatomy of Envy," in *Envy: Theory and Research*, ed. Richard H. Smith (Oxford: Oxford University Press, 2008), pp. 245–63.

25. S. E. Hill and D. M. Buss, "Envy and Positional Bias in the Evolutionary Psychology of Management," *Managerial and Decision Economics* 27 (2006): 131–43.

26. A common phenomenon stemmed from British biologist A. J. Bateman's original discovery in fruit flies in 1948: males produce more offspring when they are allowed to mate with more females whereas females produce more offspring when they are given more food.

27. See M. K. Duffy, J. D., Shaw, and J. M Schaubroeck, "Envy in Organizational Life," in *Envy: Theory and Research*, ed. Richard H. Smith (Oxford: Oxford University Press, 2008), pp. 167–89.

28. In their well-researched book, *The Spirit Level: Why Greater Equality Makes Societies Stronger* (London: Bloomsbury, 2010), K. Pickett and R. Wilkinson document statistics showing that it is society-wide equality rather than absolute personal income that underlies a nation's

health and happiness. Unsurprisingly, the United States, though ranked among the very top in absolute average income, drops below many industrial nations in a slew of key measures of well-being: life expectancy, obesity, physical health, mental health, drug use, crime rate, teenage births, violence, community life, social trust, educational performance, working hours, etc.

29. For Socrates, envy "has been acknowledged by us to be mental pain" (Plato, *Philebus* [Charleston: BibioBazaar, 360 BCE/2007], p. 110). For Aristotle, envy "is also a disturbing pain excited by the prosperity of others" (Aristotle, *Rhetoric*, in *Basic Works of Aristotle*, ed. R. McKeon [New York: Random House, 350 BCE/1941], p. 1386).

30. R. H. Smith, D. J. Y. Combs, and S. M. Thielke, "Envy and the Challenges to Good Health," in *Envy: Theory and Research*, ed. R. H. Smith (Oxford: Oxford University Press, 2008), pp. 290–314; J. Elster, *Alchemies of the Mind: Rationality and the Emotions* (Cambridge: Cambridge University Press, 1999), p. ix (and many other places in the book).

31. See S. Bezruchka, "The Status Syndrome: How Social Standing Affects Our Health and Longevity," *New England Journal of Medicine* 352 (2005): 1159–60; S. A. Everson-Rose and T. T. Lewis, "Psychosocial Factors and Cardiovascular Diseases," *Annual Review of Public Health* 26 (2005): 469–500; Smith, Combs, and Thielke, "Envy and the Challenges to Good Health."

32. M. Stein, "'Winners' Training and Its Trouble," *Personnel Review* 29 (2000): 445–59.

33. E. Haisley, E. R. Mostafa, and G. Loewenstein, "Subjective Relative Income and Lottery Ticket Purchases," *Journal of Behavioral Decision Making* 21 (2008): 283–95.

34. R. Chadha and P. Husband, *The Cult of the Luxury Brands: Inside Asia's Love Affair with Luxury* (London: Nicholas Brealey 2006), p. 1; R. W. Belk, "Marketing and Envy," in *Envy: Theory and Research*, ed. Richard H. Smith (Oxford: Oxford University Press, 2008), pp. 211–26.

35. Frank, *Luxury Fever*, p.160.

36. Based on Maynard Keynes's view that saving is detrimental to economic growth, government economic policies may also favor spending and discourage saving in order to benefit economic expansion.

37. R. P. Vecchio, "Negative Emotion in the Workplace: Employee Jealousy and Envy," *International Journal of Stress Management* 7 (2000): 161–79; R. P Vecchio, "Explorations in Employee Envy: Feeling Envious and Feeling Envied," *Cognition and Emotion* 19 (2005): 69–81; Y. Cohen-Charash and J. Meuller, "Does Perceived Unfairness Exacerbate or Mitigate Counterproductive Work Behaviors Related to Envy?" *Journal of Applied Psychology* 92 (2007): 666–80.

38. D. M. Cowherd and D. I. Levine, "Product Quality and Pay Equity between Lower-Level Employees and Top Management: An Investigation of Distributive Justice Theory," *Administrative Science Quarterly* 37 (1992): 302–20.

39. M. Bloom, "The Performance Effects of Pay Dispersion on Individuals and Organizations," *Academy of Management Journal* 42 (1999): 25–40.

40. M. K. Duffy, J. D., Shaw, and J. M Schaubroeck, "Envy in Organizational Life," in

Envy: Theory and Research, ed. Richard H. Smith (Oxford: Oxford University Press, 2008), pp. 167–89.

41. Spinoza, *Ethics*, p. 151.

42. Actually, there is an English word for this feeling, epicaricacy, which is derived from Greek. However, it is so rarely used that scholars often forgo it for the German word, Schadenfreude.

43. R. H. Smith et al., "Envy and Schadenfreude," *Personality and Social Psychology Bulletin* 28 (1996): 158–68.

44. N. L. Brigham et al., "The Role of Invidious Comparisons and Deservingness in Sympathy and Schadenfreude," *Basic and Applied Social Psychology* 19 (1997): 363–80.

45. T. Singer et al., "Empathic Neural Responses Are Modulated by the Perceived Fairness of Others," *Nature* 439 (2006): 466–69.

46. B. Vidaillet, *Workplace Envy* (New York: Palgrave Macmillan, 2008), pp. 113–15.

47. W. G. Parrott and P. M. Mosquera, "On the Pleasures and Displeasures of Being Envied," in *Envy: Theory and Research*, ed. R. H. Smith (Oxford: Oxford University Press, 2008), pp. 117–32.

48. R. Patai, *The Arab Mind* (New York: Scribner, 1973).

49. S. C. Henegan and A. G. Bedeian, "The Perils of Success in the Workplace: Comparison Target Responses to Coworkers' Upward Comparison Threat," *Journal of Applied Social Psychology* 39 (2009): 2438–68.

50. M. K. Duffy, J. D., Shaw, and J. M Schaubroeck, "Envy in Organizational Life," in *Envy: Theory and Research*, ed. R. H. Smith (Oxford: Oxford University Press, 2008), pp. 167–89. Bonds might also be attempting to avoid being questioned about his alleged use of performance-enhancing substances.

51. I here equate envy with the feeling of injustice, as they may be emotionally indistinguishable. To a certain degree, such a feeling is justified, as the income taxes are lower for millionaires than for middle-class people in the United States, a situation that has inspired President Obama and Warren Buffett both to claim, "I am paying less [taxes] than my secretary."

52. Mark 10:25.

53. P. Lookwood and Z. Kunda, "Superstars and Me: Predicting the Impact of Role Models on the Self," *Journal of Personality and Social Psychology* 73 (1997): 91–103.

54. Confucius, *The Analetics* (New York: Penguin, 1979), p. 85 (6: 29).

55. Ibid, pp. 145–148 (17: 13, 24).

56. Ibid, p. 99 (9: 26).

Chapter 6: The Soul of Anti-intellectualism

1. Except those whose family members were engaged in some professions such as begging or prostitution.

2. A. Brown, *The Rise and Fall of Communism* (New York: HarperCollins, 2009), pp. 315–16.

3. It may be hard for some in democratic nations to fathom the level of impact a leader can have in a country where power is concentrated at the top.

4. Brown, *Rise and Fall of Communism*, pp. 326–27.

5. Z. A. Medvedev, *The Rise and Fall of T. D. Lysenko*, trans. I. M. Lerner (New York: Columbia University Press, 1969), pp. 17–18.

6. This was reflected in a short conversation between Lysenko and Vavilov (see Medvedev, *Rise and Fall of T. D. Lysenko*, pp. 60–63).

7. Medvedev, *Rise and Fall of T. D. Lysenko*, pp. 20, 22.

8. Recent studies in epigenetics show that acquired inheritance can happen, but it does not occur according to the process Lamarck originally theorized.

9. Medvedev, *Rise and Fall of T. D. Lysenko*, pp. 16–17.

10. Ibid., pp. 16, 42, 64. Lysenko and his allies cherry-picked results favorable to his theory while avoiding or suppressing unfavorable results.

11. Medvedev, *Rise and Fall of T. D. Lysenko*, p. 11.

12. Ibid., p. 3.

13. Ibid., pp. 114–17.

14. Ibid., p. 65.

15. Ibid., pp. 72–73. Vavilov's escape from the death penalty was believed to be due to Lavrentiy Beria's personal interference.

16. Ibid., p. 74.

17. Ibid., pp. 123–24.

18. R. Hofstadter, *Anti-intellectualism in American Life* (New York: Knopf, 1964), p. 7.

19. Ibid., p. 6.

20. Ibid., pp. 147–48.

21. Ibid., p. 160.

22. Mencken, *Baltimore Evening Sun*, July 26, 1920.

23. Hofstadter *Anti-intellectualism in American Life*, p. 233.

24. Ibid., p. 23.

25. Ibid., p. 24.

26. R. J. Herrnstein and C. Murray, *The Bell Curve: Intelligence and Class Structure in American Life* (New York: Free Press, 1994), pp. 59, 100.

27. Heritability of IQ is a highly controversial issue. To make test scores comparable, participants must have been raised in *exactly* the same environment. Because Herrnstein and Murray's data about IQ in blacks and whites were obtained in very different social environments, they were not comparable. In other words, the methodology was flawed.

28. There exists a large amount of research literature about the relationship between IQ (general or specific, emotional or cognitive) and career success. It seems that the popular reactions to the IQ debate in the United States, China, and many other countries may have been a little carried away. Although geniuses undeniably exist, it is too ambitious to predict a person's

future with a single indicator. Even though there is a discernible statistical trend, it may be difficult for IQ to be applied to individual cases.

29. In the United States, negative perceptions about the intellectually challenged are tangible, though not openly discussed due to political correctness. So discrimination against those who are intellectually challenged is implicit.

30. H. H. Goddard, "Mental Tests and the Immigrant," *Journal of Delinquency* 2 (1917): 243–77.

31. Paul Lombardo, "Eugenic Sterilization Laws," Image Archive on the American Eugenics Movement, http://www.eugenicsarchive.org/html/eugenics/essay8text.html (accessed March 20, 2012).

32. For details and references about McCarthyism, see "McCarthyism," *Wikipedia*, http://en.wikipedia.org/wiki/McCarthyism (accessed March 20, 2012).

33. E. Schrecker, "Comments on John Earl Haynes' 'The Cold War Debate Continues,'" *Journal of Cold War Studies* 2 (2000): 1.

34. "Qian Xuesen," *Wikipedia*, http://en.wikipedia.org/wiki/Qian_Xuesen (accessed March 10, 2012).

35. Ibid.

36. Hofstadter, *Anti-intellectualism in American Life,* p. 10.

37. Ibid., p. 13. Interestingly, such rhetoric can still be heard today. In February 2012, for example, Rick Santorum roused a crowd of over a thousand supporters in Troy, Michigan, by saying that President Obama was a snob for wanting everybody in America to go to college.

38. Ibid., p. 273.

39. Ibid., p. 354.

40. Ibid., pp. 4–5.

41. PISA 2009 Technical Report (http://nces.ed.gov/pubs2012/2012045_3.pdf [accessed January 29, 2013]).

42. US Department of Education report (http://nces.ed.gov/pubs2012/2012045_3.pdf [accessed January 29, 2013]).

43. For a detailed account on the issue, see C. B. Howley, A. Howley and E. D. Pendarvis, *Out of Our Minds: Anti-intellectualism and Talent Development in American Schooling* (New York: Columbia University Press, 1995).

44. Here is a simple example to illustrate how unfriendly the American educational environment can be for children who do well in school. When my son was in third grade, there was a poster on the wall with a cute Labrador puppy holding a roll of newspaper in its mouth. Across the top, the caption read, "It's Okay to Be Smart." Few of us would take this as encouragement for excellence in learning. On the contrary, the message in the poster conveyed exactly the opposite!

45. J. Weisberg, *The Bush Tragedy* (New York: Random House, 2008), p. xix.

46. Strictly speaking, Bush's first presidency was decided by the Supreme Court.

47. Weisberg, *The Bush Tragedy*, p. xxiv. It appears he also criticizes all activists, aids, and political donors who helped get Bush elected.

48. S. Levitt, "Testing Theories of Discrimination: Evidence from *The Weakest Link*," *Journal of Law and Economics* 47 (2004): 431–52. Also see K. Antonovics, P. Arcidiacono, and R. Walsh, "Games and Discrimination: Lessons from *The Weakest Link*," *Journal of Human Resources* 40 (2005): 918–47. In addition, both studies find that women tend to discriminate against men but not vice versa.

49. A feasible solution to mitigate intellectual inequality is to open other avenues for social advancement.

PART 3: JUSTICE BY ANY MEANS

Chapter 7: Massacre in a Village

1. Our discussion of this murder case, including the excerpts of the interrogation and Hu's court statement, is based on news reports published in the Chinese magazine *Nan Feng Chuang* and information from several Chinese websites. The story is summarized in "Hu Wenhai," Baidu Baike, http://baike.baidu.com/view/893763.htm (accessed July 10, 2013). For the official ruling of the case, see "The Murder Case of Hu Wenhai, Liu Haiwang, and Hu Qinghai," Criminal Ruling Document No. 23, 2002, Higher People's Court of Shanxi Province (full text available at http://www.110.com/panli/panli_10938.html [accessed July 10, 2013]).

2. P. Marongiu and G. Newman, *Vengeance: The Fight against Injustice* (Totowa, NJ: Rowman & Littlefield, 1987), p. 7.

3. Exodus 21:23–25.

4. See, for example, Marongiu and Newman, *Vengeance*, p. 22.

5. For the top three, see "Code of Hammurabi, *Wikipedia*, http://en.wikipedia.org/wiki/Code_of_Hammurabi (accessed April 14, 2012); for the bottom two, see D. Johnston, *A Brief History of Justice* (New York: Wiley-Blackwell, 2011), p. 17.

6. Cited from D. E. Mills, "Kataki-Uchi: The Practice of Blood-Revenge in Pre-modern Japan," *Modern Asian Studies* 10 (1976): 525–42.

7. In China, dynamite is sold in sticks shaped like candlesticks. Though a controlled substance, people can get it from a variety of venues.

8. Judicial decisions in China can vary greatly from region to region and be affected by many factors. The relationship between the severity of punishment and the scale of economic damage is perceivable only in a general sense.

9. S. Jacoby, *Wild Justice: The Evolution of Revenge* (New York: Harper & Row, 1983); M. Bekoff and J. Pierce, *Wild Justice: The Moral Lives of Animals* (Chicago: University of Chicago Press, 2009).

10. Mills, "Kataki-Uchi."

11. Jacoby, *Wild Justice*, p. 14.

12. Cited from M. E. McCullough, *Beyond Revenge: The Evolution of the Forgiveness Instinct* (San Francisco, Jossey-Bass, 2008), pp. 62–64.

13. M. Daly and M. Wilson, *Homicide* (New York: Aldine de Gruyter, 1988), p. 242.

14. Jacoby, *Wild Justice*, p. 126; A. A. Roth, "The Dishonor of Dueling," *Origin* 16 (1989): 3–7.

15. S. Pinker, *The Better Angels of Our Nature: Why Violence Has Declined* (New York: Viking, 2011), p. 22.

16. McCullough, *Beyond Revenge*, pp. 30–31.

17. Daly and Wilson, *Homicide*, pp. 221–51; C. E. Kubrin and R. Weitzer, "Retaliatory Homicide: Concentrated Disadvantage and Neighborhood Culture," *Social Problems* 50 (2003): 157–80; McCullough, *Beyond Revenge*, pp. 30–31.

18. Daly and Wilson, *Homicide*, p. 226.

19. For a historical review of the idea, see E. Fromm, *The Anatomy of Human Destructiveness* (New York: Holt Rinehart & Winston, 1973), pp. 13–32.

20. At the time, cars were just becoming popular in China, and many excited new owners often drove them around for fun.

21. Jacoby, *Wild Justice*, p. 149.

22. McCullough, *Beyond Revenge*, p. 25.

Chapter 8: The Origin of Wild Justice

1. Game theory is the study of conflict and cooperation using mathematical models under the assumption that players make rational, intelligent decisions out of self-interest. It is broadly used in disciplines related to animal and human behavior including economics, biology, political science, and psychology. Most games mentioned in this book—the Ultimatum, Dictator, Cooperation, and Prisoner's Dilemma Games—are among the popular ones that have been intensively studied. In a simple game involving two players, a set of strategies and their payoffs are presented for Player 1 and Player 2, respectively. In a game where one player's gain is exactly the other player's loss, we call the game zero-sum. Otherwise, it is a nonzero-sum game.

2. J. S. Minas et al., "Some Descriptive Aspects of Two-Person Non-zero-sum Games, II," *Journal of Conflict Resolution* 4 (1960): 193–97. The researchers also used several other games with a similar payoff structure. The example used here was popularized by zoologist D. P. Barash (*Survival Game: How Game Theory Explains the Biology of Cooperation and Competition* [New York: Henry Holt, 2003]).

3. Such a combination of strategies, in the lingo of game theory, is called a *Nash equilibrium*, named after John Nash, the mathematician who shared the 1994 Nobel Prize in Economics with John Harsanyi and Reinhard Selten.

4. I here use the simple difference to emphasize the gap in relative payoff (with 0 as the reference point) between the two players. In evolutionary biology, however, a relative payoff is usually calculated as the proportion of the total payoff.

5. In game theory, it is called a dominant strategy.

6. Economist T. Riechmann ("Relative Payoffs and Evolutionary Spite," Discussion Paper no. 260 [2002], University of Hannover) provides a general form of the Spite Game, which is slightly different in format. To make the payoffs easier to understand, the values in our Spite Game are assigned differently from his standard method.

7. Barash, *Survival Game*, p. 18.

8. T. Ashworth, *Trench Warfare, 1914–1918: The Live and Let Live System* (New York: Homes & Meier, 1980); R. M. Axelrod, *The Evolution of Cooperation* (New York: Basic Books, 1984), pp. 73–87. The case is commonly explained using the paradigm of the Prisoner's Dilemma with mutual defection as the dominant strategy.

9. F. Nietzsche, *On the Genealogy of Morals*, trans. W. Kaufmann and R. J. Hollingdale (New York: Random House, 1887/1967), p. 75.

10. R. W. Harris, "Aggression, Superterritories, and Reproductive Success in Tree Swallows," *Canadian Journal of Zoology* 57 (1979): 2072–78; R. Gadagkar, "Can Animals Be Spiteful?" *TREE* 8 (1993): 232–34; K. R. Foster, T. Wenseleers, and F. L. W. Ratnieks, "Spite: Hamilton's Unproven Theory," *Annales Zoologici Fennici* 38 (2001): 229–38; R. F. Inglis et al., "Spite and Virulence in the Bacterium *Pseudomonas aeruginosa*," *Proceedings of the National Academy of Sciences, USA* 106 (2009): 5703–5707. Not all researchers agree with the interpretation that these are examples of spite.

11. Axelrod, *Evolution of Cooperation*, p. 8.

12. Fortran is a powerful computer programming language popular among scientists and engineers, especially in the 1970s and 1980s.

13. M. Milinski, "TIT FOR TAT and the Evolution of Cooperation in Sticklebacks," *Nature* 325 (1987): 433–35; L. A. Dugatkin, "Dynamics of the TIT FOR TAT Strategy during Predator Inspection in the Guppy (*Poecilia reticulate*)" *Behavioral Ecology and Sociobiology* 29 (1991): 127–32.

14. Axelrod, *Evolution of Cooperation*, p. 54.

15. M. L. Butovskaya et al., "The Hormonal Basis of Reconciliation in Humans," *Journal of Physiological Anthropology and Applied Human Science* 24 (2005): 333–37.

16. T. Singer et al., "Empathic Neural Responses Are Modulated by the Perceived Fairness of Others," *Nature* 439 (2006): 466–69.

17. J. J. Ellis, *Founding Brothers: the Revolutionary Generation* (New York: Knopf, 2000), pp. 40–41.

18. Milton, *Paradise Lost*, book IX, line 171.

19. M. E. McCullough, *Beyond Revenge: The Evolution of the Forgiveness Instinct* (San Francisco, Jossey-Bass, 2008), p. 5.

20. C. V. Witvliet, T. E. Ludwig, and K. L. V. Laan, "Granting Forgiveness or Harboring Grudges: Implications for Emotion, Physiology, and Health," *Psychological Science* 12 (2001): 117–23.

21. Cited from D. P. Barash and J. E. Lipton, *Payback: Why We Retaliate, Redirect Aggression, and Take Revenge* (New York: Oxford University Press, 2011), pp. 119–20.

22. J. B. Silk, "The Form and Function of Reconciliation in Primates," *Annual Review of Anthropology* 31 (2002): 21–44; F. B. de Waal and J. J. Pokorny, "Primate Conflict and Its Relation to Human Forgiveness," in *Handbook of Forgiveness*, ed. E. L. Worthington (New York: Routledge, 2005), pp. 17–32; McCullough, *Beyond Revenge*, pp. 120–24.

23. M. Rokeach and S. J. Ball-Rokeach, "Stability and Change in American Value Priorities, 1968–1981," *American Psychologist* 44 (1989): 775–84.

24. R. L. Gorsuch and J. Y. Hao, "Forgiveness: An Exploratory Factor Analysis and Its Relationships to Religious Variables," *Review of Religious Research* 34 (1993): 333–47.

25. Cited from McCullough, *Beyond Revenge*, p. 23.

26. H. A. Jack, ed., *The Gandhi Reader: A Sourcebook of His Life and Writings* (Bloomington: Indiana University Press, 1956), p. 319.

27. Ibid., p. 345.

28. Cited from McCullough, *Beyond Revenge*, p. 221.

29. Ellis, *Founding Brothers*, p. 43. Burr was described as "concealing his motives, covering his tracks, and destroying much of his private correspondence."

30. Ibid., p. 38.

31. Ibid., p. 31.

32. Ibid., p. 38.

33. According to information released in 1992, twenty of the forty-two Soviet missiles were equipped with nuclear warheads, ready to be used if the United States invaded Cuba. Also, a total of 42,000 Soviet soldiers entered Cuba (see J. G. Castañeda, *Compañero: The Life and Death of Che Guevara*, trans. M. Castañeda [New York: Knopf, 1997], pp. 228–29).

34. See S. Jacoby, *Wild Justice: The Evolution of Revenge* (New York: Harper & Row, 1983), pp. 5, 67–69. "Avenge not yourselves, but rather give place unto wrath: for it is written, Vengeance is mine; I will repay, saith the Lord" (Romans 12:19).

35. Ellis, *Founding Brothers*, p. 39.

36. K. F. Otterbein, "An Analysis of Homicides in Eastern Kentucky in the Late Nineteenth Century," *American Anthropologist* 102 (2000): 231–43.

37. M. Gladwell, *Outliers: The Story of Success* (New York: Little Brown, 2008), p. 166. Also, the duration of the Baker-Howard feud is from the same source.

38. C. E. Kubrin and R. Weitzer, "Retaliatory Homicide: Concentrated Disadvantage and Neighborhood Culture," *Social Problems* 50 (2003): 157–80.

39. For examples from ethnic peoples in the Mediterranean, Middle East, and North Africa, see J. Black-Michard, *Cohesive Force: Feud in the Mediterranean and the Middle East* (Oxford: Basil Blackwell, 1975).

40. R. B. Felson, "Impression Management and the Escalation of Aggression and Violence," *Social Psychology Quarterly* 45 (1982): 245–54; S. H. Kim, R. H. Smith, and N. L. Brigham, "Effects of Power Imbalance and the Presence of Third Parties on Reactions to Harm: Upward and Downward Revenge," *Personality and Social Psychology Bulletin* 24 (1998): 353–61; R. Kurzban, P. DeScioli, and E. O'Brien, "Audience Effects on Moralistic Punishment," *Evolution and Human Behavior* 28 (2007): 75–84.

41. Hu's qualm about private settlement was certainly legitimate. In the 1950s, for instance, a feudal war broke out between the Falchi and the Pes clans in Sardinia. For a while, even the Italian police were unable to stop the killing (P. Marongiu and G. Newman, *Vengeance: The Fight against Injustice* [Totowa: Rowman & Littlefield, 1987], pp. 76–85). Likewise, a resumed blood feud in the Brazilian town of Exu led to nine deaths in 1981. The Roman Catholic Church intervened but failed to make the feudists reach a private settlement. The vendetta didn't end until the military police were called in to take control (see Jacoby, *Wild Justice*, p. 123).

42. United States Department of Labor, Bureau of Labor Statistics, Workplace Shootings, Fact Sheet, 2010, http://www.bls.gov/iif/oshwc/cfoi/osar0014.htm (accessed January 29, 2013).

43. D. Cullen, *Columbine* (New York: Twelve, 2008), pp. 10, 14–15.

44. Cited from McCullough, *Beyond Revenge*, p. 34.

45. "Killer's Manifesto: 'You Forced Me into a Corner,'" CNN, April 18, 2007, http://www.cnn.com/2007/US/04/18/vtech.shooting/ (accessed July 1, 2013).

46. McCullough, *Beyond Revenge*, p. 37.

47. Axelrod, *Evolution of Cooperation*, pp. 38–40.

48. M. A. Nowak and K. Sigmund, "Tit for Tat in Heterogeneous Populations," *Nature* 355 (1992): 250–53; Nowak and Sigmund, "A Strategy of Win-Stay, Lose-Shift that Outperforms Tit-for-Tat in the Prisoner's Dilemma Game," *Nature* 364 (1993): 56–58.

49. P. Grim, "The Greater Generosity of the Spatialized Prisoner's Dilemma," *Journal of Theoretical Biology* 173 (1995): 353–59; Grim, "Spatialization and Greater Generosity in the Stochastic Prisoner's Dilemma," *Biosystems* 37 (1996): 3–17.

50. Axelrod, *Evolution of Cooperation*, p. 176.

Chapter 9: Fire behind Revolutions

1. D. Priestland, *The Red Flag: A History of Communism* (New York: Grover Press, 2009), p. xxiii. Communism can be viewed as a political system, an economic system, and an ideology (A. Brown, *The Rise and Fall of Communism* [New York: HarperCollins, 2009], pp. 105–14).

2. K. Marx and F. Engels, *The Manifesto of the Communist Party* (Chicago: Kerr, 1848/1906), p. 12.

3. Priestland, *Red Flag*, p. 87.

4. Ibid., p. 210.

5. Ibid., p. 377.

6. Ibid., p. 379.

7. Ibid., p. 386.

8. This is exemplified by the United States's support for the toppling of the Guzmán government in Guatemala; not only did the United States lose the moral high ground of democracy, but it also made more enemies in the world. While many Americans believe that the Soviet Union was defeated by long-term US policies, especially under the Reagan administration, people in other nations tend to credit Gorbachev's economic and political reforms for ending the Cold War.

9. Brown, *Rise and Fall of Communism*, p. 308.

10. R. T. Gurr was not the first to conceive the idea, but he was best known for the idea because of his in-depth treatment of the issue in his book *Why Men Rebel* (Princeton, NJ: Princeton University Press, 1970).

11. Ibid., p. 64.

12. H. Heller, *The Bourgeois Revolution in France 1789–1815* (New York: Berghahn, 2006), p. 79.

13. The constitution of French society was documented in Heller (*The Bourgeois Revolution in France 1789–1815*, p. 56). Wage earners and day laborers made up 41.4 percent; middle class, 16.9 percent; clergy and nobility, 1.1 percent.

14. Cited from D. P. Bwy, "Political Instability in Latin America: The Cross-Cultural Test of a Causal Model," *Latin American Research Review* 3, no. 2 (1968): 17–66.

15. Cited from "François-Noël Babeuf," *Wikipedia*, http://en.wikipedia.org/wiki/Gracchus_Babeuf (accessed March 3, 2012) and Brown, *Rise and Fall of Communism*, p. 16.

16. Priestland, *Red Flag*, p. 3.

17. See Bwy, "Political Instability in Latin America."

18. Japan is an exception. Its economic expansion didn't result in widening inequality in the 1970s and 1980s. Not surprisingly, it has become one of the most stable and peaceful societies in the world, despite its horrific militarism and violence in Asia during World War II.

19. For a close examination of gross inequality as a destabilizing factor in relation to the demise of empires and states in major civilizations, see M. I. Midlarsky, *The Evolution of Inequality: War, State Survival, and Democracy in Comparative Perspective* (Stanford: Stanford University Press, 1999), pp. 89–147.

20. A. de Tocqueville, *Democracy in America*, (New York: Schocken Books, 1835/1961), 2: 302.

21. Vandalism, for instance, is an expression of anger when a society is perceived as unfair and unjust. Not surprisingly, vandals mainly come from those at the bottom of society. It's hard to imagine middle- or upper-class people engaging in such activities often. Vandalism shows that when people feel victimized by unfair social or economic conditions, they are increasingly likely to disregard their social responsibilities or even turn against society.

22. See, for example, G. S. Becker, "Crime and Punishment: An Economic Approach,"

Journal of Political Economy 76 (1967): 169–17; L. Sigelman and M. Simpson, "A Cross-National Test of the Linkage between Economic Inequality and Political Violence," *Journal of Conflict Resolution* 21 (1977): 105–28; B. de Mesquita, "The War Trap Revisited," *American Political Science Review* 79 (1985): 156–77; E. N. Muller and M. A. Seligson, "Inequality and Insurgency," *American Political Science Review* 81 (1987): 425–52; M. I. Midlarsky, "Rulers and the Ruled: Patterned Inequality and the Onset of Mass Political Violence," *American Political Science Review* 82 (1988): 491–509; M. L. Lichbach, "An Evaluation of 'Does Economic Inequality Breed Political Conflict?'" *World Politics* 41 (1989): 431–70; C. D. Brockett, "Measuring Political Violence and Land Inequality in Central America," *American Political Science Review* 86 (1992): 169–76; H. Binswanger, K. Deininger and G. Feder, "Power, Distortions, Revolt and Reform in Agricultural Land Relations," in *Handbook of Development Economics*, vol. 3B, ed. J. Behrman and T. N. Srinivasan (Amsterdam: Elsevier Science, 1993), pp. 2659–72; K. Schock, "A Conjectural Model of Political Conflict: The Impact of Political Opportunities on the Relationship between Economic Inequality and Violent Political Conflict," *Journal of Conflict Resolution* 40 (1996): 98–133; F. Stewart, *The Root Causes of Conflict: Some Conclusions*, Queen Elisabeth House, Working Paper Series, no. 16 (1998); W. Nafziger and J. Auvinen, "Economic Development, Inequality, War, and State Violence," *World Development* 30 (2002): 153–63; K. Abbink, D. Masclet, and D. Mirza, "Inequality and Riots: Experimental Evidence," CREED Working Paper (2009), available at https://www.gate.cnrs.fr/IMG/pdf/Masclet.pdf (accessed May 5, 2012).

23. E. N. Muller, "Income Inequality, Regime Repressiveness, and Political Violence," *American Sociological Review* 50 (1985): 47–61.

24. M. L. Besançon, "Inequality in Ethnic Wars, Revolutions, and Genocides," *Journal of Peace Research* 42, no. 4 (2005): 393–415. The Gini index is a common measure of economic inequality. It ranges from 0, when everybody is equal, to 1, when one person has all the wealth of a society. A Gini index above 0.4 is generally considered dangerous to social stability. The following are some income Gini indices for the decade 2000–2010: the United States, 0.41; Canada, 0.33; Germany, 0.28; Japan, 0.24; China, 0.42; Mexico, 0.52; Namibia, 0.74 (see United Nations Development Programme, *The Real Wealth of Nations: Pathways to Human Development, Human Development Report 2010* [New York: Palgrave Macmillan, 2010], http://hdr.undp.org/en/media/HDR_2010_EN_Complete_reprint.pdf [accessed February 1, 2013]).

25. See, for example, J. A. Goldstone, "Toward a Fourth Generation of Revolutionary Theory," *Annual Review of Political Science* 4 (2001): 139–87.

26. I. Kawachi, B. Kennedy, and R. Wilkinson, "Crime: Social Disorganization and Relative Deprivation," *Social Science and Medicine* 48 (1999): 719–31. The results have been repeatedly confirmed in other studies (e.g., M. Daly and M. Wilson, "Risk-Taking, Intrasexual Competition, and Homicide," *Nebraska Symposium on Motivation* 47: 1–36).

27. See J. M. Cruz et al., "La Violencia en El Salvador en los Años Noventa. Magnitud, Costos y Factores Posibilitadores," Inter-American Development Bank, Working Paper R-338 (1998); P. Fajnzylber, D. Lederman and N. Loayza, "Determinants of Crime Rates in Latin

America and the World: An Empirical Assessment," World Bank Latin American and Caribbean Studies. Washington, DC: World Bank (1998); F. Stewart, "The Root Causes of Conflict: Some Conclusions," Queen Elisabeth House, Working Paper Series no. 16 (1998); R. Briceño-Leon et al., "La Violencia en Venezuela: Dimensionamiento y Políticas de Control," Inter-American Development Bank, Working Paper R-373 (1999). The Robin Hood Index is similar to the more commonly used Gini Index. In both indices, a higher value indicates a more unequal distribution of wealth and other resources.

28. Brown, *Rise and Fall of Communism*, p. 42.

29. Muller, "Income Inequality, Regime Repressiveness, and Political Violence."

30. D. McCullough, *1776* (New York: Simon & Schuster, 2005), p. 158.

31. Lichbach, "An Evaluation of 'Does Economic Inequality Breed Political Conflict?'"

32. Cited from Brown, *Rise and Fall of Communism*, p. 108.

33. Priestland, *Red Flag*, p. 2. S. Fleischacker (*A Short History of Distributive Justice* [Cambridge, MA: Harvard University Press, 2004], pp. 42–43) finds three common themes in thought regarding communal ownership of property, such as that of Plato, More, and the Jewish and Christian sacred texts: "(1) religious reasons for suspicion of wealth, (2) a notion that significant inequalities in wealth breed disharmony in society, are a source of crime and rebellion, and (3) a belief that great gaps between rich and poor citizens make it more likely that political power and economic power will become identical, that law will be used to serve the interests of the wealthy rather than to further the common good. All three . . . were to have great impact on proposals for redistributing wealth in the medieval and early modern periods. They gave rise to three fairly distinct traditions of egalitarianism."

34. Priestland, *Red Flag*, p. 114.

35. Another factor for the triumph of the Chinese Communist revolution was the utter failure of the Nationalist financial system. Inflation from 1945 to 1949 skyrocketed, which eventually led to the collapse of the currency (Brown, *Rise and Fall of Communism*, p. 183). This led to a loss of control by the Nationalist government.

36. Brown, *Rise and Fall of Communism*, pp. 228, 332–33.

37. Priestland, *Red Flag*, p. 573.

38. Brown, *Rise and Fall of Communism*, pp. 48, 50.

39. Lenin, *The State and Revolution* (Moscow: Foreign Languages Publishing House, 1917/1962), p. 145.

40. Cited from S. E. Hanson, *Time and Revolution: Marxism and the Design of Soviet Institutions* (Chapel Hill: University of North Carolina Press, 1996), p. 43.

41. Priestland, *Red Flag*, p. 571.

42. P. J. Dosal, *Comandante Che: Guerrilla Soldier, Commander, and Strategist, 1956–1967* (University Park: Pennsylvania State University Press, 2003), p. 212.

43. Cited from J. G. Castañeda, *Compañero: The Life and Death of Che Guevara*, trans. M. Castañeda (New York: Knopf, 1997), p. 186.

44. Priestland, *Red Flag*, pp. 562–63, 433.

45. Ibid., pp. 156, 159, 162, 164, 170.

46. J. K. Fairbank, *The United States and China* (Cambridge, MA: Harvard University Press, 1983), p. 427.

47. Brown, *Rise and Fall of Communism*, p. 193.

48. Ibid., p. 57.

49. Cited from Priestland, *Red Flag*, p. 179.

50. M. Perrie, ed., *The Cambridge History of Russia: The Twentieth Century* (New York: Cambridge University Press, 2006), 3: 40.

51. K. S. Tsai, *Capitalists without Democracy: The Private Sector in Contemporary China* (Ithaca, NY: Cornell University Press, 2007), p. 45.

52. Brown, *Rise and Fall of Communism,* p. 606.

53. Ibid., p. 604.

54. Cited from J. L. Anderson, *Che Guevara: A Revolutionary Life* (New York: Grove Press, 1997), p. 468.

55. Priestland, *Red Flag*, p. 385.

56. China News Agency (Zhongxinshe), April 4, 2001.

57. S. Chen and M. Ravallion, "The Developing World Is Poorer Than We Thought, But No Less Successful in the Fight against Poverty," World Bank Report (August 2008).

58. H. Beech, "Big Brotherhood," *Time*, October 22, 2012, pp. 36–42.

59. Priestland, *Red Flag*, pp. 568–69.

60. "Che Guevara Photographer Dies," BBC News, May 26, 2001, http://webcache.googleusercontent.com/search?q=cache:http://news.bbc.co.uk/2/hi/americas/1352650.stm (accessed July 2, 2013).

61. Priestland, *Red Flag*, p. 402.

62. Ibid., p. 575.

63. US Bureau of Justice Statistics, NCJ 231681, 2010, http://bjs.ojp.usdoj.gov/content/pub/pdf/cpus09.pdf (accessed January 29, 2013).

64. See, for example, H. Chenery et al., *Redistribution with Growth: Policies to Improve Income Distribution in Developing Countries in the Context of Economic Growth* (London: Oxford University Press, 1974); R. Bénabou, "Inequality and Growth," National Bureau of Economic Research, Working Paper no. 2123 (1996); T. Killick, "Responding to Inequality," Inequality Briefing Paper no. 3, DFID and ODI (2002); P. Justino, "Redistribution and Civil Unrest," paper presented at the American Economic Association meeting, Philadelphia, 2005, https://aeaweb.org/assa/2005/0107_1430_0403.pdf (accessed March 29, 2013).

65. For a detailed account, see J. E. Stiglitz, *The Price of Inequality: How Today's Divided Society Endangers Our Future* (New York: Norton, 2012).

Chapter 10: In the Mind of Terror

1. The story of the Dubrovka Theater hostage crisis was synthesized from many sources, including J. B. Dunlop (*The 2002 Dubrovka and 2004 Beslan Hostage Crises: A Critique of Russian Counter-terrorism* [Ibidem-Verlag, 2006]) and the HBO documentary *Muslim Terror in Moscow*.

2. See, for example, J. Victoroff, "The Mind of the Terrorist: A Review and Critique of Psychological Approaches," *Journal of Conflict Resolution* 49 (2005): 3–42; J. J. F. Forest, "Exploring Root Causes of Terrorism: An Introduction," in *The Making of a Terrorist: Recruitment, Training, and Root Causes*, ed. J. J. F. Forest (Westport, CT: Praeger Security International, 2006), pp. 1–14.

3. P. Wilkinson, *Terrorism versus Democracy: The Liberal State Response* (London: Routledge, 2006), p. 7. "Such justification [for terrorism] has typically rested on lofty collectivistic ideologies involving justice to the 'people,' freedom from oppression, service to God, or retribution for crimes against one's nation," write A. Kruglanski and S. Fishman in "The Psychology of Terrorism: 'Syndrome' Versus 'Tool' Perspectives," *Terrorism and Political Violence* 18 (2006): 193–215.

4. The letter can be found in many places on the Internet. It is also quoted in such published sources as M. E. McCullough, *Beyond Revenge: The Evolution of the Forgiveness Instinct* (San Francisco: Jossey-Bass, 2008), p. 38. A. J. Marsella ("Reflections on International Terrorism: Issues, Concepts, and Directions," in *Understanding Terrorism: Psychological Roots, consequences, and Interventions*, ed. F. M. Moghaddam and A. J. Marsella [Washington, DC: American Psychological Association, 2004], pp. 3–47) also shows that revenge sits atop the list of motives for anti-American terrorism.

5. B. S. Lambeth, "Russia's Air War in Chechnya," *Studies in Conflict and Terrorism* 19 (1996): 365–84.

6. See, for example, A. Speckhard and K. Akhmedova, "The Making of a Martyr: Chechen Suicide Terrorism," *Studies in Conflict and Terrorism* 29 (2006): 1–65.

7. Before they were defeated by the Spanish conquistadores, the Incas used Quechua, an oral language that was also used to "record" Inca history and culture through a few people designated as "rememberers." The Spanish conquistadores, while suppressing the practices of Inca culture, "rewrote" local history by installing their versions on those rememberers. As a result, details about the brutalities of the conquistadores were largely forgotten and the Inca's will for revenge subsided over time, and much of Inca history has been lost for good.

8. Cited from Marsella, "Reflections on International Terrorism."

9. Human Rights Watch, *Under Orders: War Crimes in Kosovo* (New York: Human Rights Watch, 2001), p. 25.

10. M. Keen, *The Outlaws of Medieval Legend* (London: Routledge and Kegan Paul, 1977), p. xix.

11. When an absolute authority such as the state cannot assume its judicial power for ret-

ribution, revenge-motivated violence will rise. Lack of such an impartial "monopoly power" to end spite is a main reason for international conflicts.

12. One difficulty in international legislation and law enforcement is that powerful nations tend to defy international laws and treaties when the pursuit of their own interests is encumbered (remember the Bush doctrine?). This reminds us of Churchill's scathing quote (chapter 2) about human history.

13. Cited from Forest, "Exploring Root Causes of Terrorism."

14. See, for example, A. P. Schmid, "Goals and Objectives of International Terrorism," in *Current Perspectives on International Terrorism*, ed. R. O. Slater and M. Stohl (New York: St. Martin's Press, 1998), pp. 47–87.

15. Kruglanski and Fishman, in "The Psychology of Terrorism," write, "The emerging consensus is that no systematic personality differences seem to demarcate terrorists as a category from non-terrorists."

16. Ibid. For this reason, terrorist organizations such as Hamas often adjust their actions according to the needs, demands, and sentiments of the communities that support them.

17. Martha Graybow and Daniel Trotta, "Bernard Madoff's Elder Son Dead in Suicide," Reuters, December 11, 2010, http://www.reuters.com/article/2010/12/11/us-madoff-son-suicide-idUSTRE6BA1GE20101211 (accessed June 15, 2012).

18. Z. Mao, *Selected Works of Mao Zedong* (Beijing: People Publishing House, 1991), 1: 1.

19. W. M. Kelley et al., "Finding the Self? An Event Related fMRI Study," *Journal of Cognitive Neuroscience* 14 (2002): 785–94; J. P. Mitchell, C. N. Macrae, and M. R. Banaji, "Dissociable Medial Prefrontal Contributions to Judgments of Similar and Dissimilar Others," *Neuron* 50 (2006): 655–63.

20. See, for example, M. A. Hogg and D. Abrams, *Social Identifications: A Social Psychology of Intergroup Relations and Group Processes* (London: Routledge, 1990); M. B. Brewer, "Ingroup Identification and Intergroup Conflict: When Does Ingroup Love Become Outgroup Hate?" in *Social Identity, Intergroup Conflict, and Conflict Resolution,* ed. R. D. Ashmore, L. Jussim, and D. Wilder (New York: Oxford University Press, 2001), pp. 17–41.

21. D. P. Barash and J. E. Lipton, *Payback: Why We Retaliate, Redirect Aggression, and Take Revenge* (New York: Oxford University Press, 2011), p.139.

22. A. Kohut, "Support for Terror Wanes among Muslim Publics," Pew Global Attitudes Project (2005).

23. S. Levin et al., "Social Dominance and Social Identity in Lebanon: Implications for Support of Violence against the West," *Group Processes & Intergroup Relations* 6 (2003): 353–68.

24. N. J. Smelser, *The Faces of Terrorism: Social and Psychological Dimensions* (Princeton, NJ: Princeton University Press, 2007), p. 68.

25. Cited from Wilkinson, *Terrorism versus Democracy*, p. 40.

26. National Commission on Terrorist Attacks, *The 9/11 Report* (New York: St. Martin's Press, 2004), p. 76.

27. Smelser, *Faces of Terrorism*, p. 110.

28. See Victoroff, "Mind of the Terrorist."

29. Smelser, *Faces of Terrorism*, p. 148.

30. Keen, *Outlaws of Medieval Legend*, pp. 217–18.

31. P. Tyler, *A World of Trouble: The White House and the Middle East—from the Cold War to the War on Terror* (New York: Farrar, Straus & Giroux, 2009), pp. 14–15. In his address to the US Congress on September 20, 2001, President Bush, when answering his own question, "Why do they [al-Qaeda] hate us?" answered with a non sequitur, "They hate our freedoms—our freedoms of religion, our freedom of speech, our freedom to vote and assemble and disagree with each other."

32. P. R. Pillar, "Superpower Foreign Policies: A Source for Global Resentment," in *The Making of a Terrorist: Recruitment, Training, and Root Causes*, ed. J. J. F. Forest (Westport, CT: Praeger Security International, 2006), pp. 31–44.

33. Holding it as a state secret, the Russian government has never released the true identity of the mysterious gas.

34. For a detailed account, see Dunlop, *The 2002 Dubrovka and 2004 Beslan Hostage Crises*.

35. Smelser, *Faces of Terrorism*, p. 197.

36. Wilkinson, *Terrorism versus Democracy*, p. 91.

37. M. Sageman, *Understanding Terror Networks* (Philadelphia: University of Pennsylvania Press, 2004), pp. 111–13. Kinship plays a considerable role in the recruitment of terrorists in the Basque independence movement, al-Qaeda, and the Palestine Liberation Organization. See D. T. Schiller, "A Battlegroup Divided: The Palestinian Fedayeen," in *Inside Terrorist Organizations*, ed. D. C. Rapoport (London: Frank Cass, 2001), pp. 90–108; J. M. Post, "Killing in the Name of God: Osama bin Laden and Al Qaeda," in *Know Thy Enemy: Profiles of Adversary Leaders and Their Strategic Cultures*, ed. B. Schneider and J. M. Post (Maxwell Air Force Base, AL: USAF Counter-Proliferation Center, 2003), pp. 17–40; Smelser, *Faces of Terrorism*, p. 98.

38. E. H. Kaplan et al., "What Happened to Suicide Bombings in Israel? Insights from a Terror Stock Model," *Studies in Conflict & Terrorism* 28 (2005): 225–35.

39. Based on his speech on September 22, 2003 (see Forest, "Exploring Root Causes of Terrorism").

40. Marsella, "Reflections on International Terrorism."

41. See "George W. Bush," *Wikipedia*, http://en.wikiquote.org/wiki/George_W._Bush (accessed June 20, 2012).

42. A major reason behind the Bush administration's willingness to enter the war was to "spread democracy," conceived and promoted by Bush's neoconservative advisors and supporters.

43. Quoted in Art Moore, "Democratic Senator Praises Bin Laden," WND, December 20, 2002, http://www.wnd.com/2002/12/16360/ (accessed July 10, 2013).

44. See Austin Bay, "What the US Needs Is a Liberal Hawk," Strategy Page, January 8, 2003, http://www.strategypage.com/on_point/20030108.aspx (accessed July 10, 2013).

45. Smelser, *Faces of Terrorism*, pp. 86–87. During the G. W. Bush presidency, Joe Biden, then a US senator, proposed the construction of a thousand schools in Afghanistan to counter radical religious teaching there. The project was quickly killed because "spending twenty million dollars on schools in Afghanistan is a harder sell than spending four hundred billion on defense; fear is more compelling than foresight" (cited from K. von Hippel, "Dealing with the Roots of Terror," in *The Making of a Terrorist: Recruitment, Training, and Root Causes*, ed. J. J. F. Forest [Westport, CT: Praeger Security International, 2006], pp. 266–76).

46. Pillar, "Superpower Foreign Policies." Instances of international strife have also been attributed to envy (S. Schimmel, *Seven Deadly Sins* (New York: Bantam Doubleday, 1993). Also, F. Zakaria ("The Politics of Rage," *Newsweek,* October 15, 2001, pp. 22–40) ascribes global Islamic terrorism to envy and resentment for Western power and influence. Anti-Semitism in Nazi Germany is a historical example of envious prejudice (P. Glick, "Sacrificial Lambs Dressed in Wolves Clothing: Envious Prejudice, Ideology, and the Scapegoating of Jews," in *What Social Psychology Can Tell Us about the Holocaust*, ed. L. S. Newman and R. Erber [Oxford: Oxford University Press, 2002], pp. 113–42).

47. Marsella, "Reflections on International Terrorism."

48. Forest, "Exploring Root Causes of Terrorism."

49. See M. T. Klare, "Fueling the Fires: The Oil Factor in Middle Eastern Terrorism," in *The Making of a Terrorist: Recruitment, Training, and Root Causes*, ed. J. J. F. Forest (Westport, CT: Praeger Security International, 2006), pp. 140–59; K. Von Hippel, "Dealing with the Roots of Terror."

50. Pillar, "Superpower Foreign Policies." Smelser (*The Faces of Terrorism*, p. 215) presents a similar view: "The United States is regarded as hypocritical in its pronouncements of high democratic ideas and its history of support of some brutal and totalitarian regimes."

51. A. Krueger and J. Malečková, "Does Poverty Cause Terrorism?" *New Republic*, June 24, 2002, pp. 27–33.

52. See T. L. Friedman, *Longitudes and Attitudes: Exploring the World after September 11* (New York: Farrar Straus & Giroux, 2002), p. 356.

53. See, for example, Abi–Hashem, "Peace and War in the Middle East: A Psychological and Sociocultural Perspective," in *Understanding Terrorism: Psychological Roots, Consequences, and Interventions*, ed. F. M. Moghaddam and A. J. Marsella (Washington, DC: American Psychological Association, 2004), pp. 69–89; Smelser, *Faces of Terrorism*, p. 215.

54. Cited from Pillar, "Superpower Foreign Policies."

55. R. Wright, *Rock the Casbah: Rage and Rebellion across the Islamic World* (New York: Simon & Schuster, 2012), p. 3.

56. Ibid., p. 46.

57. Ibid., p. 2.

58. Ibid., p. 3.

59. According to Wright (p. 52), "By 2010, long before bin Laden was killed in a US

raid, public opinion polls in major Muslim countries showed dramatic declines in support for al Qaeda. Support for bin Laden had dropped to 2 percent in Lebanon and 3 percent in Turkey."

60. Smelser, *Faces of Terrorism*, pp. 52–53.

PART 4: EQUALITY AND DIVINE JUSTICE

Chapter 11: The War on Polygyny

1. For a detailed account of the YFZ raid, see J. D. Weaver, "The Texas Mis-Step: Why the Largest Child Removal in Modern U.S. History Failed," *William & Mary Journal of Women and the Law* 16 (2009): 449–535. Also see A. M. Guiora, "Protecting the Unprotected: Religious Extremism and Child Endangerment," *Journal of Law & Family Studies* 12 (2010): 391–407 (and citations within).

2. See Weaver, "Texas Mis-Step."

3. For a summary of the story, see Guiora, "Protecting the Unprotected."

4. See K. Driggs, "Who Shall Raise the Children? Vera Black and the Rights of Polygamous Utah Parents," *Utah Historical Quarterly* 60 (1992): 27–46; S. M. Sigman, "Everything Lawyers Know about Polygamy Is Wrong," *Cornell Journal of Law and Public Policy* 16 (2006): 101–85.

5. J. Dougherty and K. Johnson, "Leader of Polygamist Sect Guilty in Rape Case," *New York Times*, September 25, 2007, http://www.nytimes.com/2007/09/25/us/25cnd-jeffs.html (accessed March 4, 2013). Due to a technical issue, Jeffs's conviction was reversed by the Utah's Supreme Court on July 27, 2010.

6. See "YFZ Ranch," *Wikipedia*, http://en.wikipedia.org/wiki/YFZ_Ranch (accessed September 18, 2012).

7. Cited from Sigman, "Everything Lawyers Know about Polygamy Is Wrong."

8. A. D. Davis, "Regulating Polygamy: Intimacy, Default Rules, and Bargaining for Equality," *Columbia Law Review* 110 (2010): 1955–2046.

9. Cited from Sigman, "Everything Lawyers Know about Polygamy Is Wrong."

10. M. Strassberg, "The Crime of Polygamy," *Temple Political and Civil Rights Law Review* 12 (2002/2003): 353.

11. J. Hartung, "Polygyny and Inheritance of Wealth," *Current Anthropology* 23 (1982): 1–12.

12. See D. P. Barash and J. E. Lipton, *The Myth of Monogamy: Fidelity and Infidelity in Animals and People* (New York: Holt, 2001), pp. 3–4.

13. See, for example, R. A. Posner, *Sex and Reason* (Cambridge, MA: Harvard University Press, 1992), p. 260; W. Scheidel, "Monogamy and Polygyny in Greece, Rome, and World History," Princeton/Stanford Working Papers in Classics (2008).

14. J. L. Embry, *Mormon Polygamous Families: Life in the Principle* (Salt Lake City: Kofford Books, 2008), pp. 1–3.

15. P. L. Kilbride, *Plural Marriage for Our Times: A Reinvented Option?* (Westport, CT: Bergin and Garvey, 1994), p. 53.

16. Cf. Embry, *Mormon Polygamous Families*, p. 3.

17. Scheidel, "Monogamy and Polygyny in Greece, Rome, and World History."

18. See Sigman, "Everything Lawyers Know about Polygamy Is Wrong."

19. See Kilbride, *Plural Marriage for Our Times*, pp. 58, 64; J. Cairncross, *After Polygamy Was Made a Sin: The Social History of Christian Polygamy* (London: Routledge & Kegan Paul, 1974), p. 60; "The Council of Trent, the Twenty-Fourth Session," Hanover Historical Texts Project," http://history.hanover.edu/texts/trent/ct24.html (assessed March 22, 2013).

20. Cairncross, *After Polygamy Was Made a Sin*, p. 2.

21. Cf. Embry, *Mormon Polygamous Families*, p. 4.

22. Ibid., p. 4.

23. Cf. Sigman, "Everything Lawyers Know about Polygamy Is Wrong."

24. M. Doepke and M. Tertilt, "Women's Liberation: What's in It for Men?" IZA Discussion Paper no. 3421 (2008).

25. The practice of polygyny in the Mormon community was not explicit until 1852.

26. J. Krakauer, *Under the Banner of Heaven: A Story of Violent Faith* (New York: Doubleday, 2003), pp. 6, 125.

27. Ibid., p. 117.

28. Ibid., p. 118.

29. Embry, *Mormon Polygamous Families*, pp. 6–7.

30. Ibid., p. 7.

31. Krakauer, *Under the Banner of Heaven*, pp. 125–26.

32. Mormonism should be viewed as part of Christianity. Here I separate the two words, Mormon and Christian, both for the ease of writing and for the historical fact that many Americans did not accept Mormons as Christians in the beginning.

33. Krakauer, *Under the Banner of Heaven*, p. 106. Smith declared his candidacy for president of the United States in January 1844.

34. Ibid., p. 128.

35. L. Forster, *Religion and Sexuality: Three American Communal Experiments of the Nineteenth Century* (New York: Oxford University Press, 1981), p. 124.

36. Krakauer, *Under the Banner of Heaven*, pp. 196–201.

37. Ibid., p. 204.

38. Ibid., p. 203.

39. It is interesting that Mormons were consistently against slavery whereas many Christians in the Midwest still embraced it after the Missouri Compromise.

40. Sigman, "Everything Lawyers Know about Polygamy Is Wrong."

41. Krakauer *Under the Banner of Heaven*, p. 6.

42. Sigman, "Everything Lawyers Know about Polygamy is Wrong."

43. The wording differs somewhat in different sources (e.g., Krakauer, *Under the Banner of Heaven*, p. 230).

44. N. F. Scott, *Public Vows: A History of Marriage and the Nation* (Cambridge, MA: Harvard University Press, 2000), p. 113; Sigman, "Everything Lawyers Know about Polygamy is Wrong"; E. K. Phipps, "Marriage and Redemption: Mormon Polygamy in the Congressional Imagination, 1862–1887," *Virginia Law Review* 95 (2009): 435–87.

45. M. M. Ertman, "Race Treason: The Untold Story of America's Ban on Polygamy," *Columbia Journal of Gender and Law* 19 (2010): 287–366.

46. Sigman, "Everything Lawyers Know about Polygamy Is Wrong."

47. Ibid. Also see Scott, *Public Vows*, pp. 111–13.

48. Sigman, "Everything Lawyers Know about Polygamy is Wrong."

49. Cited from Scott, *Public Vows*, pp. 111–12.

50. Sigman, "Everything Lawyers Know about Polygamy Is Wrong"; K. R. Schwab, "Lost Children: The Abuse and Neglect of Minors in Polygamous Communities of North America," *Cardozo Journal of Law & Gender* 16 (2010): 315–41.

51. Sigman, "Everything Lawyers Know about Polygamy Is Wrong."

52. Cited from Scott, *Public Vows*, p. 119.

53. Information about US government actions against polygyny was based on Embry (*Mormon Polygamous Families*, pp. 9–11).

54. Cited from Scott, *Public Vows*, p. 139.

55. According to historian J. Gray (*Rebellions and Revolutions: China from the 1800s to 2000* [Oxford: Oxford University Press, 2002], p. 62), "[Hong] insisted on the right of all men, not just the emperor, to worship Heaven, and he preached the brotherhood of man on the grounds that all men (and women) were God's children. He extended this idea of brotherhood to people of all nations. His God was a God of equality, freedom, and universal love."

56. For a detailed account of the Taiping Rebellion, see Gray, *Rebellions and Revolutions*, pp. 52–76.

57. Some historians believe he died of illness, but others think he committed suicide when Nanjing fell to the Qing armies.

58. S. H. Chang and L. H. D. Gordon, *All under Heaven: Sun Yat-sen and His Revolutionary Thought* (Stanford: Hoover Institution Press, 1991), p. 7.

59. Ibid., pp. 7–9; M.-C. Bergère, *Sun Yat-sen*, trans., J. Lloyd (Stanford: Stanford University Press, 1994), pp. 25–28.

60. Chang and Gordon, *All under Heaven*, p. 12.

61. Bergère, *Sun Yat-sen*, p. 33.

62. Ibid., pp. 249–54.

63. Chang and Gordon, *All under Heaven*, p. 149.

64. Ibid., p. 149.

65. J. Taylor, *The Generalissimo: Chiang Kai-shek and the Struggle for Modern China* (Cambridge, MA: Harvard University Press, 2009), pp. 38–40.

66. Ibid., pp. 73–76.

67. Ibid., p. 91.

68. Bergère, *Sun Yat-sen*, p. 231.

69. F. Engels, *The Origin of the Family, Private Property and the State* (Chicago: Kerr, 1902), p. 57.

70. Note that Engels (*The Origin of the Family, Private Property and the State*, p. 87) defines monogamy in a slightly different way. He writes, "the family of the proletarian is no longer strictly monogamous . . . . The proletarian marriage is monogamous in the etymological sense of the word, but by no means in a historical sense."

71. Scheidel, "Monogamy and Polygyny in Greece, Rome, and World History."

72. M. K. Zeitzen, *Polygamy: A Cross-Cultural Analysis* (Oxford: Berg, 2008), pp. 37, 147 (and the citation of J. van Wing within).

73. K. Mann, "The Historical Roots and Cultural Logic of Outside Marriage in Colonial Lagos," in *Nuptiality in Sub-Saharan Africa: Contemporary Anthropological and Demographic Perspectives*, ed. C. Bledsoe and G. Pison (Oxford: Clarendon, 1994), pp. 167–93.

74. Zeitzen, *Polygamy*, p. 147.

75. Ibid, pp. 155–56.

76. Ibid, p. 149.

77. R. Lesthaege, "On the Adaptation of Sub-Saharan Systems of Reproduction," in *The State of Population Theory: Forward from Malthus*, ed. D. Coleman and R. Schofield (Oxford: Basil Blackwell, 1986), pp. 212–38.

Chapter 12: Reproductive Fairness and the Triumph of Monogamy

1. J. Goody, *The Development of the Family and Marriage in Europe* (Cambridge: Cambridge University Press, 1983), p. 64.

2. J. Goody, *Production and Reproduction: A Comparative Study of the Domestic Domain* (Cambridge: Cambridge University Press, 1976), p. 93.

3. A similar point is also made by P. L. Kilbride, *Plural Marriage for Our Times: A Reinvented Option?* (Westport, CT: Bergin and Garvey, 1994), p. 58.

4. N.-P. Lagerlöf, "Sex, Equality, and Growth," *Canadian Journal of Economics* 38 (2005): 807–31.

5. E. D. Gould, O. Moav, and A. Simhon, "The Mystery of Monogamy," *American Economic Review* 98 (2008): 333–57.

6. S. H. Citci, "Development and Transition to Monogamy," Access Econ, March 19, 2010, http://www.econ.boun.edu.tr/pet10/PET10_Papers/PET10-10-00456.pdf (accessed March 4, 2013).

7. H. Fisher (*Anatomy of Love: Natural History of Monogamy, Adultery and Divorce* [New York: Simon & Schuster, 1993], p. 68) makes a similar point: "If polygyny were permitted in New York, Chicago, or Los Angeles, an Episcopalian man with $200 million could probably also attract several women willing to share his love—and his cash."

8. Ibid., p. 66.

9. M. Borgerhoff Mulder, "Women's Strategies in Polygynous Marriage: Kipsigis, Datoga, and Other East African Cases," *Human Nature* 3 (1991): 45–70.

10. Lagerlöf "Sex, Equality, and Growth."

11. L. L. Betzig, *Despotism and Differential Reproduction: A Darwinian View of History* (New York: Aldine, 1986); Betzig, "Sex, Succession, and Stratification in the First Six Civilizations," in *Social Stratification and Socioeconomic Inequality*, vol. 1, *A Comparative Biosocial Analysis*, ed. L. Ellis (Westport, CT: Praeger, 1993), pp. 37–74.

12. Betzig, "British Polygyny," in *Human Biology and History*, ed. M. Smith (London: Taylor and Francis, 2002), p. 85.

13. M. K. Zeitzen, *Polygamy: A Cross-Cultural Analysis* (Oxford: Berg, 2008), p. 53.

14. Betzig, "Roman Polygyny," *Ethology and Sociobiology* 13 (1992): 309–49; Betzig, "Medieval Monogamy," *Journal of Family History* 20 (1995): 181–215.

15. J. Cairncross, *After Polygamy Was Made a Sin: The Social History of Christian Polygamy* (London: Routledge & Kegan Paul, 1974), p. 70.

16. L. Stone, *The Family, Sex, and Marriage in England, 1500–1800* (New York: Harper & Row, 1979), p. 663.

17. See Zeitzen, *Polygamy*, pp. 147–48.

18. See R. Hager and C. B. Jones, ed., *Reproductive Skew in Vertebrates: Proximate and Ultimate Causes* (Cambridge: Cambridge University Press, 2009) for a general pattern in reproductive skew in animals from insects to humans.

19. Zeitzen, *Polygamy*, p. 64.

20. Fisher, *Anatomy of Love*, p. 69.

21. T. A. Telford, "Fertility and Population Growth in the Lineages of Tongcheng County, 1520–1661," in *Chinese Historical Microdemography*, ed. S. Harrell (Berkeley: University of California Press, 1995), pp. 48–93.

22. J. Chamie, "Polygyny among Arabs," *Population Studies* 40 (1986): 55–66.

23. S. M. Sigman, "Everything Lawyers Know about Polygamy is Wrong," *Cornell Journal of Law and Public Policy* 16 (2006): 101–85.

24. G. S. Becker, *A Treatise on the Family* (Cambridge, MA: Harvard University Press, 1991). Also see Gould, Moav, and Simhon, "Mystery of Monogamy."

25. Becker, "A Theory of Marriage, Part II," *Journal of Political Economy* 82 (1974): S1–S26.

26. R. Wright, *The Moral Animal: Why We Are the Way We Are—The New Science of Evolutionary Psychology* (New York: Pantheon, 1994), p. 96.

27. D. M. Buss, "Sex Differences in Human Mate Preferences: Evolutionary Hypotheses

Testing in 37 Cultures," *Behavioral and Brain Sciences* 12 (1989): 1–49; Buss, *The Evolution of Desire: Strategies of Human Mating* (New York: Basic Books, 1994); B. J. Ellis, "The Evolution of Sexual Attraction: Evaluative Mechanisms in Women," in *The Adapted Mind: Evolutionary Psychology and the Generation of Culture*, ed. J. H. Barkow, L. Cosmides, and J. Tooby (Oxford: Oxford University Press, 1992), pp. 267–88.

28. R. A. Posner, *Sex and Reason* (Cambridge, MA: Harvard University Press, 1992), pp. 253–60.

29. S. Kanazawa and M. C. Still, "Why Monogamy?" *Social Forces* 78 (1999): 25–50.

30. C. A. Hornung, "Status Relationships in Marriage: Risk Factors in Spouse Abuse," *Journal of Marriage and the Family* 43 (1981): 675–92.

31. T. Bereczkei and A. Csanaky, "Mate Choice, Marital Success, and Reproduction in a Modern Society," *Ethology & Sociobiology* 17 (1996): 17–35.

32. W. Scheidel, "Monogamy and Polygyny in Greece, Rome, and World History," Princeton/Stanford Working Papers in Classics (2008).

33. For instance, Kipsigis of Kenya (Borgerhoff Mulder, "Women's Strategies in Polygynous Marriage"), Taita of Kenya (Kilbride, *Plural Marriage for Our Times*, p. 35), and the Ivory Coast (E. Boserup, *Women's Role in Economic Development* [London: Earthscan, 1989], p. 43).

34. See K. R. Schwab, "Lost Children: The Abuse and Neglect of Minors in Polygamous Communities of North America," *Cardozo Journal of Law & Gender* 16 (2010): 315–41.

35. Cited from A. D. Davis, "Regulating Polygamy: Intimacy, Default Rules, and Bargaining for Equality," *Columbia Law Review* 110 (2010): 1955–2046.

36. See, for example, E. Porter, *The Price of Everything: Solving the Mystery of Why We Pay What We Do* (New York: Portfolio/Penguin, 2011), pp. 84–85.

37. N. Barber, "Explaining Cross-National Differences in Polygyny Intensity: Resource-Defense, Sex Ratio, and Infectious Diseases," *Cross-Cultural Research* 42 (2008): 103–17.

38. See Kanazawa and Still, "Why Monogamy?"

39. M. Tertilt, "Polygyny, Fertility, and Savings," *Journal of Political Economy* 113 (2005): 1341–71.

40. Zeitzen, *Polygamy*, pp. 92–93.

41. See Kilbride, *Plural Marriage for Our Times*, p. 77.

42. A. Wolf and C. Huang, *Marriage and Adoption in China, 1845–1945* (Stanford, CA: Stanford University Press, 1980). J. Goody (*The Oriental, the Ancient and the Primitive* [Cambridge: Cambridge University Press, 1990], p. 107) also reported a similar figure in Fujian.

43. Cited from R. White, "Two Sides of Polygamy," *Utah Law Review* 11 (2009): 495–502.

44. See A. M. Guiora, "Protecting the Unprotected: Religious Extremism and Child Endangerment," *Journal of Law & Family Studies* 12 (2010): 391–407.

45. Ibid. See also, E. Eckholm, "Boys Cast Out by Polygamists Find Help," *New York Times*, September 9, 2007.

46. For more details about "Lost Boys," see J. Borger, "The Lost Boys, Thrown Out of

US Sect so that Older Men Can Marry More Wives," *Guardian*, June 14, 2005, http://www.guardian.co.uk/world/2005/jun/14/usa.julianborger (accessed July 9, 2013).

47. See Sigman, "Everything Lawyers Know about Polygamy Is Wrong"; Guiora, "Protecting the Unprotected."

48. M. Tertilt, "Polygyny, Women's Rights, and Development," *Journal of the European Economic Association* 4 (2006): 523–30.

49. Porter (*The Price of Everything*, p. 80) also makes a similar point. N.-P. Lagerlöf ("Sex, Equality, and Growth") shows a similarity in reproductive inequality across marriage systems: "in polygynous societies it is rich and powerful men who have more wives and thus more offspring; in monogamous societies the rich and powerful tend to father more illegitimate children than poor men."

50. Genesis 9:1.

51. Jeremiah 20:30.

52. N. Barber, *Kindness in a Cruel World: The Evolution of Altruism* (Amherst, NY: Prometheus Books, 2004), p. 223.

53. F. van Balen, T. Trimbos-Kemper, and J. Verdurmen, "Perception of Diagnosis and Openness of Patients about Infertility," *Patient Education and Counseling* 28 (1996): 247–52; L. Schmidt, "Social and Psychological Consequences of Infertility and Assisted Reproduction—What Are the Research Priorities?" *Human Fertility* 12 (2009): 14–20; N. Scheper-Hughes, *Saints, Scholars, and Schizophrenics: Mental Illness in Rural Ireland* (Berkeley: University of California Press 1979).

54. F. van Balen and H. M. W. Bos, "The Social and Cultural Consequences of Being Childless in Poor-Resource Areas," *OBGYN Monograph* (2010): 1–16.

55. T. R. Malthus, *An Essay on the Principle of Population* (London: Ward & Lock, 1798/1890), pp. xii, 9.

56. J. G. Kennedy, *Herbert Spencer* (Boston: Twayne, 1978), p. 104.

57. Childless couples often find it difficult to communicate with friends who have children (see van Balen, Trimbos-Kemper, and Verdurmen, "Perception of Diagnosis and Openness of Patients about Infertility" and Schmidt, "Social and Psychological Consequences of Infertility and Assisted Reproduction—What Are the Research Priorities?").

58. Bill Maher, *Real Time with Bill Maher*, episode 232, November 11, 2011 (http://www.hbo.com/real-time-with-bill-maher/episodes/0/232-episode/article/new-rules.html, accessed July 11, 2013).

59. Porter, *The Price of Everything*, p. 83.

60. Scheidel "Monogamy and Polygyny in Greece, Rome, and World History."

61. See Fisher, *Anatomy of Love*, p. 100; Sigman "Everything Lawyers Know about Polygamy Is Wrong."

62. See Tertilt, "Polygyny, Fertility, and Savings."

63. S. T. Emlen, "The Evolutionary Study of Human Family Systems," *Social Science Information* 36 (1997): 563–89.

64. T. Flanagan, "Polygamy Is a Deterrent to Democracy," In *Polygamy*, ed. R. D. Landford Jr. (Farmington Hills, MI: Greenhaven Press, 2009).

65. A. Korotayev and D. Bondarenko, "Polygyny and Democracy: A Cross-Cultural Comparison," *Cross-Cultural Research* 34 (2000): 190–208.

66. Scheidel "Monogamy and Polygyny in Greece, Rome, and World History."

67. Cited from Scheidel "Monogamy and Polygyny in Greece, Rome, and World History."

68. R. D. Alexander et al., "Sexual Dimorphisms and Breeding Systems in Pinnipeds, Ungulates, Primates, and Humans," in *Evolutionary Biology and Human Social Behavior: An Anthropological Perspective*, ed. N. A. Chagnon and W. Irons (North Scituate, MA: Duxbury Press, 1979), pp. 402–35.

69. See H. Krahn, T. F. Hartnagel, and J. W. Gartrell, "Income Inequality and Homicide Rates: Cross-National Data and Criminological Theories," *Criminology* 24 (1986): 269–94; C.-C. Hsieh and M. D. Pugh, "Poverty, Income Inequality, and Violent Crime: A Meta-Analysis of Recent Aggregate Data Studies," *Criminal Justice Review* 18 (1993): 182–202; Kaplan et al., "Inequality in Income and Mortality in the United States: Analysis of Mortality and Potential Pathways," *British Medical Journal* 312 (1996): 999–1003; B. P. Kennedy, I. Kawachi, and D. Prothrow-Stith, "Income Distribution and Mortality: Test of the Robin Hood Index in the United States," *British Medical Journal* 312 (1996): 1004–07; W. A. Naudé, Evolutionary Psychology and Development Economics, Paper for the UNU-WIDER Jubilee Conference on the Future of Development Economics, Helsinki, Finland (2005).

70. See, for example, M. K. Bacon, I. L. Child, and H. Barry, "A Cross-Cultural Study of Correlates of Crime," *Journal of Abnormal and Social Psychology* 66 (1963): 291–300; M. Wilson and M. Daly, "Competitiveness, Risk Taking, and Violence: The Young Male Syndrome," *Ethology and Sociobiology* 6 (1985): 59–73; S. Kanazawa and M. C. Still, "Why Men Commit Crimes," *Sociological Theory* 18 (2000): 434–47; S. Kanazawa, "Why Productivity Fades with Age: The Crime-Genius Connection," *Journal of Research in Personality* 37 (2003): 257–72.

71. M. Daly and M. Wilson, *Homicide* (New York: Aldine de Gruyter, 1988).

72. D. P. Barash, "Evolution, Males, and Violence," *Chronicle of Higher Education*, May 24, 2002, http://www.physics.ohio-state.edu/~wilkins/writing/Assign/so/male-violence.html (accessed April 9, 2013).

73. R. Alexander, "The Search for a General Theory of Behavior," *Behavioral Science* 20 (1975): 77–100; Alexander, *The Biology of the Moral Systems* (New York: Aldine de Gruyter, 1987), p. 71.

74. R. Sampson, J. Laub, and C. Wimer, "Does Marriage Reduce Crimes? A Counterfactual Approach to Within-Individual Causal Effects," *Criminology* 44 (2006): 465–509.

75. Cited from J. Henrich, R. Boyd, and P. J. Richerson, "The Puzzle of Monogamous Marriage," *Philosophical Transactions of the Royal Society* B 367 (2012): 657–69.

76. J. Dreze and R. Khera, "Crime, Gender, and Society in India: Insights from Homicide Data," *Population and Development Review* 26 (2000): 335–52.

77. M. Ridley, *The Red Queen: Sex and the Evolution of Human Nature* (New York: Harper Perennial, 1993), pp. 98, 199.

78. Wright, *Moral Animal*, p. 98.

79. Betzig, "Medieval Monogamy."

80. Sigman, "Everything Lawyers Know about Polygamy Is Wrong."

81. Lagerlöf, "Pacifying Monogamy," *Journal of Economic Growth* 15 (2010): 235–62.

82. D. P. Barash and J. E. Lipton, *The Myth of Monogamy: Fidelity and Infidelity in Animals and People* (New York: Holt, 2001), p. 134.

83. See Davis, "Regulating Polygamy." Davis also cites serious studies that have been commissioned by the Canadian government to examine the feasibility of polygyny. Even in some rural areas of Russia some have lobbied and campaigned for the legal recognition of polygyny.

84. Cited from Davis, "Regulating Polygamy."

85. J. Krakauer, *Under the Banner of Heaven: A Story of Violent Faith* (New York: Doubleday, 2003), p. 252.

86. See Sigman, "Everything Lawyers Know about Polygamy Is Wrong"; Guiora, "Protecting the Unprotected."

87. Guiora, "Protecting the Unprotected."

88. According to Zeitzen (*Polygamy*, p. 89), about forty thousand to sixty thousand Mormons continue to practice plural marriage today.

89. Figure 1 is drawn from data presented in R. Stark, *Rise of Mormonism* (New York: Columbia University Press, 2005), p. 141.

Chapter 13: Comeuppance and the Higher Robin Hood

1. Although evidence is still inadequate, some scholars believe that the behavior of "rain dancing" observed in chimpanzees may be best explained as a proto-religious ritual.

2. E. O. Wilson, *Consilience: The Unity of Knowledge* (London: Abacus, 1999), p. 292.

3. E. Durkheim, *The Elementary Forms of the Religious Life*, trans., C. Cosman (Oxford: Oxford University Press, 1912/2001), p. 46.

4. This idea is known as group selection, which is almost anathema to mainstream evolutionary biologists, who generally believe that selection among groups is far too weak to be accountable when compared with selection at the individual level. Models using game theory show that conflicts between groups can promote ingroup cooperation (e.g., J.-K. Choi and S. Bowles, "The Coevolution of Parochial Altruism and War," *Science* 318 [2007]: 636–40). There is little empirical evidence showing that selective results at the group level cannot also be explained by selection at the individual level.

5. See M. J. Rossano, *Supernatural Selection: How Religion Evolved* (Oxford: Oxford University Press, 2010), pp. 167–68.

6. See H. Gintis et al., "Explaining Altruistic Behavior in Humans," *Evolution and*

Human Behavior 24 (2003): 153–72; F. L. Roes and M. Raymond, "Belief in Moralizing Gods," *Evolution and Human Behavior* 24 (2003): 126–35. See also J. Diamond's book, *Collapse* (2005), which provides a detailed, in-depth analysis of why some societies failed while others flourished.

7. A. Norenzayan and A. F. Shariff, "The Origin and Evolution of Religious Prosociality," *Science* 322 (2008): 58–62.

8. G. P. Murdock, *Theories of Illness: A World Survey* (Pittsburgh: University of Pittsburgh Press, 1980), pp. 17–27.

9. J. Haidt, "The New Synthesis in Moral Psychology," *Science* 316 (2007): 998–1002.

10. N. P. Azari, J. Missimer, and R. J. Seitz, "Religious Experience and Emotion: Evidence for Distinctive Cognitive Neural Patterns," *International Journal for the Psychology of Religion* 15 (2005): 263–81.

11. A. B. Newberg and E. G. d'Aquili, "The Neuropsychology of Spiritual Experience," in *Handbook of Religion and Mental Health*, ed. H. G. Koenig (San Diego, CA: Academic Press, 1998), pp. 75–94; A. B. Newberg et al., "The Measurement of Regional Cerebral Blood Flow during the Complex Cognitive Task of Meditation: A Preliminary SPECT Study," *Psychiatric Research: Neuroimaging* 106 (2001): 113–22; W. H. McNeill, *Keeping Together in Time: Dance and Drill in Human History* (Cambridge, MA: Harvard University Press, 1995), pp. 69–78.

12. M. Beauregard and V. Paquette, "Neural Correlates of a Mystical Experience in Carmelite Nuns," *Neuroscience Letters* 405 (2006): 186–90.

13. See, for example, B. King-Casas et al., "Getting to Know You: Reputation and Trust in a Two-Person Economic Exchange," *Science* 308 (2005): 78–83; U. Schjoedt, "Evolutionary Psychology, Neuroscience and the Study of Religion," in *The Evolution of Religion: Studies, Theories, and Critiques*, ed. J. Bulbulia et al. (Santa Margarita, CA: Collins Foundation Press, 2008), pp. 303–11; U. Schjoedt, "The Religious Brain: A General Introduction to the Experimental Neuroscience of Religion," *Method and Theory in the Study of Religion* 21 (2009): 310–39.

14. See, for example, S. P. Morgan, "A Research Note on Religion and Morality: Are Religious People Nice People?" *Social Forces* 61 (1983): 683–92; M. J. Donahue and P. L. Benson, "Religion and the Well-Being of Adolescents," *Journal of Social Issues* 51 (1995): 145–160; K. S. Kendler, C. O. Gardner, and C. A. Prescott, "Clarifying the Relationship between Religiosity and Psychiatric Illness: The Impact of Covariates and the Specificity of Buffering Effects," *Twin Research* 2 (1999): 137–44; J. L. Furrow, P. E. King, and K. White, "Religion and Positive Youth Development: Identity, Meaning, and Prosocial Concerns," *Applied Developmental Science* 8 (2004): 17–26. As a result, religious people are often viewed as more trustworthy and cooperative, though evidence indicates that religious faith plays no role in people's goodness (See D. C. Dennett, *Breaking the Spell: Religion as a Natural Phenomenon* [New York: Penguin, 2006], pp. 279–80). But the false impression has its social impact. A recent poll shows that while 60 percent of Americans approve of having a religious president, only 37 percent would be comfortable with an atheist president, compared with 38 percent who would be comfortable with a Muslim as president (Jonathan Weisman, "Poll: Voters Wary

of a Mormon President," http://blogs.wsj.com/washwire/2011/06/08/poll-voters-wary-of-a-mormon-president/ [accessed January 31, 2013]), despite sporadic anti-Muslim incidents such as burning the Quran and condemning Islam for terrorist attacks. Ironically, a recent survey shows that atheists, on average, have the best understanding of world religions, including religious tenets, main religious figures, and religious texts.

15. W. Irons, "Morality, Religion, and Human Nature," in *Religion and Science: History, Method, and Dialogue*, ed. W. M. Richardson and W. Wildman (New York: Routledge, 1996), pp. 375–400.

16. A. Greif, "Contract Enforceability and Economic Institutions in Early Trade: The Maghribi Traders' Coalition," *American Economic Review* 83 (1993): 525–48; J. Ensminger "Transaction Costs and Islam: Explaining Conversion in Africa," *Journal of Institutional and Theoretical Economics* 153 (1997): 4–29.

17. See, for example, J. H. W. Tan and C. Vogel, "Religion and Trust: An Experimental Study," European University Viadrina, Department of Business Administration Discussion Paper No. 240 (2005); Norenzayan and Shariff, "Origin and Evolution of Religious Prosociality."

18. R. Sosis and C. Alcorta, "Signaling, Solidarity, and the Sacred: The Evolution of Religious Behavior," *Evolutionary Anthropology* 12 (2003): 264–74.

19. See Haidt, "New Synthesis in Moral Psychology."

20. See H. B. Earhart, ed., *Religious Traditions of the World* (New York: HarperCollins, 1993).

21. D. D. P. Johnson and O. Krüger, "The Good of Wrath: Supernatural Punishment and the Evolution of Cooperation," *Political Theology* 5 (2004): 159–76.

22. Murdock, *Theories of Illness*, pp. 17–27.

23. W. Flesch, *Comeuppance: Costly Signaling, Altruistic Punishment, and Other Biological Components of Fiction* (Cambridge, MA: Harvard University Press, 2007), p. 90.

24. See P. McNamara, *The Neuroscience of Religious Experience* (Cambridge: Cambridge University Press, 2009), pp. 193, 195.

25. Whenever implicit in examples and metaphors, I here follow the convention adopted by Western scholars and writers, who tend to address religious issues from the perspective of the monotheistic Judeo-Christian tradition.

26. D. Ariely, *Predictably Irrational: The Hidden Forces That Shape Our Decisions* (New York: HarperCollins, 2008), pp. 283–84.

27. J. M. Bering, "The Folk Psychology of Souls," *Behavioral and Brain Science* 29 (2006): 453–98.

28. M. Foucault, *Discipline and Punish: The Birth of the Prison* (New York: Pantheon Books, 1977).

29. D. Kahneman, *Thinking, Fast and Slow* (New York: Farrar, Straus and Giroux, 2011), pp. 51–53.

30. See W. I. Thomas and D. S. Thomas, *The Child in America: Behavior Problems and*

Programs (New York: Knopf, 1928), pp. 571–72. The Thomas theorem states that "if men define situations as real, they are real in their consequences."

31. Cited from Johnson and Krüger, "Good of Wrath."

32. M. D. Alicke and E. Zell, "Social Comparison and Envy," in *Envy: Theory and Research*, ed. R. H. Smith (Oxford: Oxford University Press, 2008), pp. 73–93.

33. Cited from Rossano, *Supernatural Selection*, p. 33.

34. For a recent ranking of professionals for their honesty and ethical standards, see "Gallup Poll Least Trusted Professionals," SodaHead, http://www.sodahead.com/fun/gallup-poll-shows-congress-to-be-least-trusted-occupation/blog212277/?link=ibaf&q=gallup20poll%20least%20trusted%20professionals%202011 (accessed April 18, 2013).

35. In 1930, British biologist R. A. Fisher proposed the runaway selection hypothesis to account for the rapid evolution of traits, such as the peacock's tail, that bestow no apparent survival advantage. Despite some indirect empirical evidence, it has been largely untested.

36. See Sosis and Alcorta, "Signaling, Solidarity, and the Sacred."

37. Paraphrased from Luke 18:18–30.

38. L. R. Iannaccone, "Sacrifice and Stigma: Reducing Free-Riding in Cults, Communes, and Other Collectives," *Journal of Political Economy* 100 (1992): 271–91; Norenzayan and Shariff, "Origin and Evolution of Religious Prosociality."

39. R. Sosis and E. Bressler, "Cooperation and Commune Longevity: A Test of the Costly Signaling Theory of Religion," *Cross-Cultural Research* 37 (2003): 211–39.

40. Rossano, *Supernatural Selection*, p. 164.

41. See F. Young, *Initiation Ceremonies* (New York: Bobbs-Merrill, 1965); A. Glucklich, *Sacred Pain* (New York: Oxford University Press, 2001); Sosis and Bressler, "Cooperation and Commune Longevity."

42. From a historical point of view, despite St. Paul's advocacy for ascetics, it was not until the eleventh century that celibacy was officially imposed on all clergy. Meanwhile, abstinence from sex was used as a chief criterion to judge a priest's piety. Adultery, for both men and women, became a major sin (see H. Fisher, *Anatomy of Love: Natural History of Monogamy, Adultery and Divorce* [New York: Simon & Schuster, 1993], p. 84). For celibacy as a costly signal, see M.-A. Benz, E. Franck, and U. Meister, "Strategic Choice of Celibacy in the Catholic Church," Institute for Strategy and Business Economics, University of Zurich, Working Paper no. 42 (2005); J. P. Schloss, "He Who Laughs Best: Involuntary Religious Affect as a Solution to Recursive Cooperative Defection," in *Evolution of Religion: Studies, Theories, and Critiques*, ed. J. Bulbia et al. (Santa Margarita, CA: Collins Foundation Press, 2008), pp. 197–207; M. J. Murray and L. Moore, "Costly Signaling and the Origin of Religion," *Journal of Cognition and Culture* 9 (2009): 225–45.

43. G. E. Swanson, *The Birth of the Gods: The Origin of Primitive Beliefs* (Ann Arbor: University of Michigan Press, 1960), pp. 153–74.

44. See "List of National Anthems," *Wikipedia*, http://en.wikipedia.org/wiki/List_of_national_anthems (accessed January 12, 2013).

45. The US Constitution, though lacking any explicit reference to God, implicitly refers to the divine in several places.

46. P. L. Kilbride, *Plural Marriage for Our Times: A Reinvented Option?* (Westport, CT: Bergin and Garvey, 1994), p. 81.

47. A. E. McGrath, *Christianity's Dangerous Idea: The Protestant Revolution—A History from the Sixteenth Century to the Twenty-First* (New York: HarperOne, 2007), p. 134.

48. Ibid., pp. 139, 140, 145.

49. G. Wills, *Head and Heart: American Christianities* (New York: Penguin, 2007), pp. 478–79.

50. Ibid., pp. 467–70.

51. R. Marius, *Martin Luther: The Christian between God and Death* (Cambridge, MA: Harvard University Press, 1999), p. 6.

52. Ibid., p. 109.

53. McGrath, *Christianity's Dangerous Idea*, p. 22.

54. Marius, *Martin Luther*, p. 82.

55. Ibid., p. 39.

56. McGrath, *Christianity's Dangerous Idea*, p. 3.

57. Marius, *Martin Luther*, p. 146.

58. Ibid., p.370.

59. Ibid.

60. H. C. Kee et al., *Christianity: A Social and Cultural History* (New York: Macmillan, 1991), p. 344.

61. Marius, *Martin Luther*, pp. 157–58.

62. Kee et al., *Christianity*, p. 347.

63. Marius, *Martin Luther*, p. 235.

64. McGrath, *Christianity's Dangerous Idea*, p. 25.

65. Ibid., p. 7.

66. Rossano, *Supernatural Selection*, p. 210.

67. Cited from Johnson and Krüger, "Good of Wrath."

EPILOGUE: THE FUTURE OF FAIRNESS

1. P. Corning, *The Fair Society: The Science of Human Nature and the Pursuit of Social Justice* (Chicago: University of Chicago Press, 2012), p. 165.

2. Winston Churchill, speech in the House of Commons, November 11, 1947; see Answers, http://wiki.answers.com/Q/Who_said_democracy_is_the_worst_form_of_government (accessed July 10, 2013).

3. See David P. Barash, "Unequal or Unfair? Which Is Worse: Inequality Is a Symptom; Unfairness Is the Disease," *Pura Vida* [blog], November 13, 2012, http://www.psychologytoday.com/blog/pura-vida/201211/unequal-or-unfair-which-is-worse (accessed February 24, 2013).

INDEX